A Modern
DRY-FLY
Code

VINCENT MARINARO

WITH A NEW FOREWORD BY DATUS C. PROPER

ILLUSTRATIONS BY PEARCE BATES

THE LYONS PRESS
New York

Printed in the United States of America.

10 9 8 7 6 5 4 3 2 1

Library of Congress Cataloging-in-Publication Data

Marinaro, Vincent.
A modern dry-fly code / Vincent Marinaro; with a new afterword by
Datus C. Proper; illustrations by Pearce Bates.
p. cm.
Originally published: New York: Putnam, c1950.
ISBN 1-55821-413-5
1. Fly Fishing. 2. Trout Fishing—Pennsylvania. 3. Aquatic insects—Pennsylvania.
4. Insects—Pennsylvania. I. Title.
SH456.M3 1997
799.1'757—dc21   97-29512
CIP

Contents

FOREWORD

A MODERN DRY-FLY CODE opened a half-century during which anglers would proliferate, equip themselves amply, and absorb great blocks of Vincent Marinaro's original thinking into conventional wisdom. And yet after all that, his work still seems innovative. You read it today and find that life is just catching up.

In 1950, when the first edition was published, American fly fishing remained intensely (because unwittingly) provincial. Every dry fly for trout sold by a major mail-order firm[1] still followed Frederic Halford's 1886 British designs. Patterns with names like Iron Blue, Blue Dun, and Cowdung imitated insects that do not exist in America. There was one unlikely caddisfly in the catalogue and one possible stonefly. There were no spent spinners, no terrestrial imitations, no hooks smaller than size 14, and not a single fly likely to deceive a trout in the Letort Spring Run.

It was not that American anglers made a point of copying British flies. We had, on the contrary, lost track of whom we were copying. We were taking the advice of writers who had borrowed from other writers and so on back to 1886, when somebody read Halford. We were working from copies of copies of copies, and the product had lost its relation to nature—American or British. Halford's original Iron Blue Dun might have worked on the

---

[1] The Orvis catalogue, 1959—nine years after Marinaro's first edition.

Letort, if only because it was small, but an Iron Blue on a size 14 Mustad hook (equivalent to a Redditch 12) would have put down rising brown trout in chalkstream and spring creek alike.

At core, the problem was that fishing authorities, with honorable exceptions, were dispensing advice uncluttered by sources. It was an old habit among writers on both sides of the Atlantic, and I mention it here because Vince Marinaro raised the subject with me. He had acknowledged his own debts and was not amused when his personal contributions were later borrowed without attribution.

Vince returned to original sources, natural and human. He collected local stream insects, had them identified by entomologists, and rethought the artificial fly from head to tail. Earlier American writers were of little help in this work because none had understood the limestone spring creeks. Marinaro referred to predecessors "like Hewitt and La Branche and Gill" as "legendary."

In British books, however, Vince found what he needed—not on specific insects but on methods. He opened the *Code* with a quotation from Col. E.W. Harding, then drew from Skues, Halford, Mottram, Dunne, Ronalds, and more. Finding the right sources must have taken research, for an American in the 1940s.

A Modern Dry-Fly Code was not the first American work on flies that imitate natural insects. Jennings and Flick had both published before 1950; both knew Catskill trout and mayflies[2]; and both (in my opinion) tied excellent dry flies in the traditional design. This, however, was a subject on which Vince

---

[2] Note for British readers: A mayfly, to anglers in America (and scientists everywhere), is any member of the order of Ephemeroptera.

did not agree, as I learned when he went through a manuscript of my first book. The Halfordian (and Catskill) dry fly was, for him, merely a wet fly adapted to float—a purpose for which the design was not suited. With this background you will understand Marinaro's meaning when, in the pages that follow, he regrets that G.E.M. Skues did not "emancipate" the floating fly as he did the wet.

Marinaro (unlike Skues) left no list of angling contributions. Lest we forget, consider some innovations in the *Code.*

1. **Terrestrial flies** This book gave land-based insects their myth—and some of their best designs. There is a floating ant with hackle in the center of the body, a brilliant jassid, ingenious beetles, and a grasshopper unlike any other.

2. **Minutae** Americans often need smaller flies than British anglers, but we did not know that till the *Code* taught us so.

3. **Widespread tail** Marinaro was, I think, first to describe "the enormous mechanical advantages to be gained by a proper arrangement of tail fibres" in the dry fly. A divided tail helps in persuading a winged fly to land and float in the correct position. This idea (with variations in the method of tying) has been widely adopted since 1950.

4. **The "thorax" fly** Hackles are wound well back from the eye of the hook—an idea for which the author gave credit to Edgar Burke—and designed to make the fly float flat or slightly nose-down on the water, like a real mayfly. Marinaro's original design is still used, though it is not easy to tie. Many successors use other approaches to the same end.

5. **Olives** Marinaro may have been first to recognize the importance of mayflies in the genus *Baetis* on American waters. (In 1969, he would also alert anglers to the genus *Tricorythodes*. Taken together, olives and tricos now furnish more than half of my fishing with imitative flies.)

In all of the above, what matters is not the author's specific patterns or tying methods, which can be altered to suit each individual fly tyer. What matters is discovery.

There was yet another discovery, if one uses the term in a sense made popular by European explorers of new lands. Marinaro put limestone spring creeks on the American angler's map. It required a "brand of fly-fishing . . . never observed or exploited before my time," he writes. He must have worked out the chalkstream method by reading, for he would not fish the River Itchen until years later.

There are spring creeks west of the Great Plains which are, today, in better condition than either the Pennsylvania limestoners or the English chalkstreams. There are tailwater fisheries that provide the same kind of fishing, and more of it, without sources in springs. The American fly-fishing boom of recent years has focused on such fertile streams. In them we catch rising fish, or try to catch them, by matching the hatch. It involves stalking a visible quarry, rather than waiting for something mysterious to happen in the depths. The people who are drawn to fly fishing in the first place are often especially drawn to this particular kind—but we were not aware of that before 1950.

Few people have proved more thoroughly than Vincent Marinaro that fly fishing is an intellectual passion. He taught himself to make horsehair lines, using an authentic gadget found at a flea market. He reconstructed the old British North-Country flies, taking pains to find authentic materials. (Who else had dotterel feathers?)

His passions had nothing to do with price or prestige. I heard him express admiration for a few books, a cock's cape with silver-colored hackles, some old Hardy silk lines, one or two Partridge hooks, good double-barreled shotguns, a rod by Tom Maxwell, and a pair of hackle pliers. "That's the only good pair of hackle pliers I ever saw," he said.

The list of things he did not like was longer but expressed with equal frankness, if one asked. He held conventional wisdom in such disregard that some interlocutors found him unsettling. In addition to Halfordian dry flies, he had no time for:

- Rivers (or grouse coverts) with lots of people in them.
- Writers who attract crowds by publicizing individual streams.
- Anglers who fail to respect their prey. "Fishing is a blood sport," he said, and certain obligations come with it.
- Some prestigious bamboo rods, especially if they had stiff butts or soft middles.
- All graphite rods. He found them lacking in soul, repulsive, "almost slimy," and got so that he would not willingly walk into a shop where he had to look at them. (But, at an earlier stage, he once admitted that an Orvis 9′3″ graphite rod for a 6-weight line cast well.)

It was easy to know when Vince was not pleased, and as the years went on, he increasingly objected to overwhelming trout with modern technology. You may be sure that I did not "pollute the water" (his term) with plastic rods when we went fishing together.

He insisted on "treating the stream right," but that did not mean putting all fish back. He liked a trout dinner, especially when it was cooked by his wife.

Once when I mentioned a well-known American author in conversation, Vince shook his head and said that he had seen the man "fishing the water" (casting at random) at a time when there were visible rises to cover. I did not know enough to press the point, but I wondered whether the other angler could see rising trout as well as Marinaro—whose eyesight was keen long after his legs gave out.

He was, however, not a dry-fly purist. On one of our last trips to the Letort, he experimented with old-fashioned wet flies on small double hooks. I don't recall seeing him nymph-fishing. One day, though, I was fishing a little herl-bodied beetle upstream and wet, just like a Skues-style nymph, and Vince invited me to try two small Letort fish that had refused his dry fly. He seemed delighted when they took the beetle. I think he enjoyed filing that away as another angling problem solved.

Marinaro made his own rods of split cane, starting with a double-handed salmon rod that seemed impossibly light. His personal favorite was a 9-foot, 3-piece, 4-ounce rod for a 6-weight line. (His bad hip made wading difficult, and the long rod kept backcasts above foliage on the bank.) My favorites were the 8- and 7½ footers, which had about two-thirds the weight of my own rods—and cast better.

Then there was the 6-foot rod for a 3-weight line. It weighed, Vince said, just under an ounce, and it was a real fishing tool, not a toy like the old Leonard "Baby Catskill." We proved this point with long, easy casts under the old apple tree in his back yard.

The rods' tapers were, for him, not a matter of individual preference: There were specific tasks that had to be performed well or the rod was simply bad. But he was uncharacteristically evasive on the particulars. If asked, he would say that he did not want anyone making bad rods from his convex tapers, and the other tricks of the trade were as important as the dimensions in thousandths of an inch. He had hoped that some rod-maker with a milling machine would ask him to put rods in production. (He used planing forms only because he had no alternative.) He would not sell individual rods because he didn't want "to sell a $10,000 design for $1,000."

I urged Vince to put what he knew about rod-building in a book, because it was difficult enough to make bamboo rods that would hold their own against the synthetics, and he should not let the best designs disappear without a ripple. He did not write the book, but I still hope that his tapers and notes will be made available to rod-builders.

For me, the watershed in American fly fishing came with 1950 and the *Code*. In the years since, other good books have appeared and vast (but still incomplete) work has been done on American trout-stream insects. There could have been no better model than Marinaro. He was a lawyer, and the *Code* made case law that would be argued before the court of anglers without proving flawed. Vince claimed nothing that he had not done. There were no evasive generalities to fail examination. Precedents were identified. If you open the book today for the first time, you will not feel that you are reading something obsolete. Everything in it works, and always will.

Datus Proper
Belgrade, MT

A BACKWARD LOOK

UPON reviewing A MODERN DRY-FLY CODE for this new edition I am struck by the remarkable intensity of the narrative, which seems a little strange to me now. For me fly-fishing was never a contemplative man's recreation. I could never achieve the placidity or the detachment expressed in Izaak Walton's *Compleat Angler;* since Walton practiced and wrote about a blood sport, he and his book will remain forever anomalies in fishing literature. If anything, fly-fishing was and is for me a constant state of excitement, and my attitude is best likened to that of a hound dog joyously baying in full, hot pursuit of its quarry. True, there are many sections in the "Code" that seem to have a philosophical flavor; but this is only because they represent some attempt to rationalize the behavior of trout and the imitations in the form of dry flies. The excitement that I mention was and is generated as much by the preliminary moves as by the ultimate capture of a trout. The mysteries of fly-fishing and the inscrutable ways of trout will forever fascinate me.

All of it began when, as a young man from the raw mountain country of western Pennsylvania, I looked for the first time on the fair face of the Letort and neighboring limestone streams of the Cumberland Valley in Pennsylvania. I knew instinctively that these waters were very different from anything that I had hitherto known. They had a rich, fertile look about them that contained the promise of great fulfillment in the way of fishing pleasure. I did not know then as I do now that many similar waters—rich, weedy, full of insect and fish life—

existed elsewhere and are particularly numerous in the Midwest and the far West. I was, perhaps, somewhat provincial in my outlook (so I have been accused), believing that nothing like my limestones existed except for the chalk streams of England and France. Anyway, I was happy to be so confined, not caring to look elsewhere, and hugely enjoying the superb, fascinating, and exciting kind of trout fishing that I found in the limestone country of central Pennsylvania. In essence, my approach to the fishing on these waters was contained in the stalking habit that I learned as a boy in the mountain country. Standing for long moments, surveying the ground or the water where the quarry lives, avoiding sudden movements, approaching slowly and carefully, and using all available cover—these are the disciplines of the stalking animal. It was and still is the only successful way to approach the wild and fearful trout of the Letort. It was and still is the only way to observe and resolve the mysteries of fly-fishing: to collect and distill all of the facts to establish sound fly-fishing practice.

Fly-fishing in its best form, in the best circumstances, is a rather secretive and solitary kind of sport. Imagine then my amazement when I saw for the first time, particularly on the opening day of the trout season, the great hordes of fishermen gathered around certain favorite spots, elbow to elbow and cheek by jowl, hurling a wonderful assortment of indelicate lures at their intended victims, the trout lying almost directly underfoot—and catching them too! Even today, after having seen this performance repeated for many years, I cannot get used to the idea that this is the way most fishermen desire to fish. It can happen only with artificially propagated, hand-fed trout, familiar with the presence of humans, coming docilely to feed at the sound of falling pellets. What a dreadful kind of deterioration for both stant companions in those days, brought together by a common bond, our total devotion to the wonderful world of trout and fly-fishing. I spent a lot of time in that hut, on weekends and vacation periods. In it there had been placed, by Bob McCafferty, a professional flytier and mutual friend, a rather large and elaborate chest containing a wealth of fly-tying materials—a really fine gift. I was privileged to use the contents of that chest as freely and as often as I liked, and I did so. In it there was a considerable amount of genuine jungle cock nail feathers in various sizes. I had no appreciable familiarity with this feather although I knew that it was extensively employed in the tying of salmon flies and streamer types of wet flies.

It was a fine arrangement. The water was only a few feet away from the hut and the fly-tying bench. The trout rose steadily to the daily fall of beetles. I had only to tie a fly, rush out, and give it an immediate trial. Constantly goaded by repeated refusals from the trout, I continued to fashion numerous unsuccessful models of the beetle. Then, one day, I performed a very simple act, stupendous in its consequences. I added one of those beautiful oval-shaped shiny-looking jungle cock nail feathers to one of my artificials in an effort to improve its conformation and to simulate better the natural insect.

As with many previous models, the fly floated badly because of the water-soaking bulky body that I persisted in using. Additionally and importantly, those bodies prevented the jungle cock from lying flat and flush with the surface of the water, in the same way that the natural insect behaved. Finally, in a fit of impatience I dispensed with the body completely, using nothing but hackle and the jungle cock nail. Thus was born not only the first successful beetle pattern but, more significantly, the fabulous little Jassid. It was only a brief step from the beetle, a rather large insect, to the smaller jassid— only a matter of size. The beetle was parent to the Jassid and grandparent to numerous new patterns effectively imitating caddis and stoneflies. The remarkable success of this style of tie, now widely acclaimed, undoubtedly accounts for the rash of "inventors" of the Jassid, offered as a great discovery under new-sounding names. Never was any bastard child more freely acknowledged by more unexpectant fathers.

The response from the trout to the new beetle and Jassid style of tie was prompt and electrifying! It was the first breakthrough in a long series of breakthroughs that opened the door to a new and utterly fascinating kind of fly-fishing. For a long time the new terrestrial patterns remained unpublicized. Fox did not tie flies in those days; he was the first and only person to whom I gave some of the new patterns with the confidence that they would get a skillful and thorough tryout. His early and immediate successes were phenomenal. Together, thereafter, we enjoyed this late-season fishing virtually alone and uninterrupted. Gradually, news of our activities leaked out and spread afar, attracting many eager fly-fishermen from near and far, particularly the more astute and appreciative kind, men such as Hewitt, LaBranche, and Joe Brooks.

The beetle is with us no more. Nevertheless, that which it inspired and that which I learned from its brief presence will remain forever valid in application to terrestrial dry flies in fly-fishing.

An overpowering curiosity led me from the beetle to other forms of land-bred insects. Very early in this strange game, I was immensely impressed by the abundance of ants, in the meadows, on the water, and as revealed by the autopsies of trout. And truthfully, if I were to choose one pattern above all others, day in and day out, from fish to fish, the most enduring in its season, it would be the ant in its various sizes and colors. Yes, I would choose it over the Jassid!

I could find nothing in American fly-tying literature to help me fashion a good pattern of the ant. It seems to have been completely ignored by American fishermen. British authors did not ignore the ant, but the tie recommended by Halford and others was unacceptable to me and to the trout. It looked exactly like any other upright-winged pattern of a mayfly. The first really sensible model of an ant that I saw was tied by the late McCafferty. He tied the ant with a large round knob near the bend of the hook, a smaller knob near the eye, a thin waist between them, and only a few hackle fibers. He never tied anything but a black pattern and always in large sizes, perhaps 10 or 12. He always tied them to sink, never to float, something which puzzled me exceedingly, for the ants do not sink but float very well, even the large carpenter ants. The two knobs and the narrow waist were the important elements of form that I adopted for my own floating patterns. Thereafter it became a matter of variation in size and color-choosing suitable body materials to make the ant float and, importantly, tying its flying or winged form for those frequent periods when the great flights sometimes blanket the water from bank to bank.

Throughout all of my adventures with the insects of my home waters, I became increasingly aware of a special class of insects quite distinct from pattern or genealogy. That class is distinctive in its very small size. Every order of insect life is represented by extremely tiny species. There are very small beetles, ants, jassids, and mayflies, to name only a few. I have created a special small-fly class because it creates special problems.

In those early days I could not take advantage of the marvelous small-fly fishing that prevailed on the limestone waters, simply because I did not have the tools for the job. I was frequently galled to the core when I found good trout feeding incessantly on miniscule insects for hours on end and I could not make a fair try for them because I did not have flies small enough or gut fine enough to handle them properly. The very small hooks in sizes 22, 24, and 28 and very fine gut in 6X, 7X, and 8X were not available. Today all of that is changed by the advent of very fine nylon tippet material and very small hooks in sizes 22, 24, and 28. But in those early days I was a prophet without honor, for I was severely challenged for even suggesting, as I did in Chapter 3, that size 14 was the largest that need be used for any dry-fly imitation. Heavens! Today fishermen are catching Atlantic salmon on size 14. Indeed, there is a cult of salmon fishermen in which membership is earned by catching a sixteen-pound salmon on a size 16 hook, called, I think, the 16–16 club!

Today fishermen are generally enjoying the superlative sport and challenge of taking big trout on tiny flies and fine gut. It has become rather commonplace. And today I am oftentimes delighted immeasurably by strangers whom I meet casually on the stream, who want to help with my fishing fortunes by offering me small flies and instructions for their use. Bless them for their kindness!

This brief review would not be complete without some reference to the mayflies or Ephemeroptera, the traditional models for imitation. In the twenty years that have passed since the initial publication of A MODERN DRY-FLY CODE, I have not seen or experienced anything to make me change my position on the scope or theory of imitation. I believed then and do now that fly-fishermen have been sorely overtaxed by the superabundance of fly patterns. Catalogs of fly patterns numbering in excess of five hundred varieties represent the accumulated debris of five centuries of fly-tying disorder. Much of this is complicated by patterns that do not expresss any theory of imitation but are designed as attractors. Selecting a reasonable number of useful patterns from such a vast array is an appalling confrontation. If you believe as I do that the bodies of the duns are meaningless and superfluous, then with one mighty stroke you have eliminated a great deal of confusion and uncertainty. You need imitate only the wing, which is of paramount importance and which is limited to very few colors: the large blue-gray group and the yellow group. I am going a bit further than one keen observer who said many years ago: "Let your imitations be of any color as long as they are gray."

Again, if you believe as I do that the body of the spinner is of paramount importance because it has broken through the opaque mirror and is fully visible to the trout, then you need to employ very few colors, and those chiefly in the thorax, for the empty abdomen is usually devoid of any pronounced color.

The principal point of departure expressed by my spinner patterns is contained in the fact that they are floated by means of the body and not by means of hackle. It has been very difficult for fishermen to accept the idea that hackle can be eliminated from a floating fly. As a matter of fact any hackle, intended as a support, presents a very distorted view of the imitation to the trout. The complete outline of the spinner—thorax, abdomen, tails, and half-spent or fully spent wings—exactly as the natural appears, is the prime objective. Whether or not you agree with my method of obtaining this effect is only incidental. Certainly for extremely large imitations, as for the Green Drake spinner, the porcupine quill pattern is the best floater that I have been able to devise. For smaller spinners, which can be tied on a smaller and lighter hook, nothing surpasses seal's fur, when it is obtainable. Only recently has some of this superb material been made available by a limited permissible harvest of seals, once threatened with extinction. I am of course speaking of the hard, brilliant, translucent type of seal's fur, not the soft, dull patches of processed scraps gathered from the furriers.

Above all else let me commend and recommend to the fly-fishermen the thorax style of tie for both duns and spinners. For me, it has been a first-rate performer, far superior to anything tied with older techniques.

This short backward look cannot and does not treat adequately the questions and omissions that have become apparent with the passage of time and the discerning scrutiny of my readers. It is possible that in the near future I may write another fishing book which will treat the omissions. Meanwhile, if the reader cannot fish with me in person, let him fish with me in this book, and together we shall contend with the lusty trout of the limestone country.

VINCENT C. MARINARO

Mechanicsburg, Pennsylvania
1970

Chapter 1

WHAT PRICE IMITATION?

"TO SOME all this may seem like taking a recreation far too seriously. If these objectors can take lightly the sense of baffled disappointment following on failure by the waterside; if they are content to enjoy success as though it were some caprice of chance; if, in short, they are content to be the slaves and not the masters of their fishing fate, then perhaps they are right: but to me the sense of bafflement robs me of half my pleasure and casual unexplained success is but Dead Sea fruit to the palate of enjoyment."

COL. E. W. HARDING

The Fly Fisher and the Trout's Point of View

THAT statement deserves the stamp of immortality. It contains the fundamental philosophy of the fly-fisherman and it expresses the best justification, if any is

3

needed, for further complicating the fascinating art of fly-fishing. Anyway, fly-fishing was never a very simple affair. From the very beginning, from Dame Juliana's time, it has been a decidedly obscure and difficult art to fathom. Certainly there is nothing simple about Dame Juliana's treatise, the first book on fly-fishing. Deception is her keynote and deception is not simplicity. Imitation is her constant theme and that requires considerable skill, artfulness, and good craftsmanship. She set the pattern of endeavor for all time. From that day onward, a host of fishermen writers, experimenters, and observers have sought and learned in a like manner, hinting, suggesting, formulating, and generally poking a tentative finger at the veil of mystery that surrounds this pleasant pastime. The general nature of these efforts does no more than scratch the surface of an impenetrable wall behind which are hidden the great mysteries of the underwater world, a world no less fantastic in actuality than Kingsley's *Water Babies* or Carroll's *Alice in Wonderland,* and all of the fishermen's accumulated knowledge of that world seems to be no more than a collection of such scratches, no matter what form of fly-fishing is involved, dry or wet.

Of course, earlier fishermen and fishermen-writers were concerned only about the wet fly and a wealth of material can be found on that subject. The dry fly, that wonderful dry fly, is a comparatively new invention needing a separate treatment, but the object of investigation is still the same, imitation and deception. It is quite unnecessary and decidedly bad taste to reopen the old argument of dry fly versus wet fly, that strange controversy, the echoes of which can still be heard now and then. Let it be generally accepted that each has its place and either of them constitutes tolerable practice wherever it is applicable. Let personal preference, *sans peur, sans reproche,* determine the final choice. Mine is the dry fly!

4

"What is the Dry?" asked Emlyn Gill in 1912, first American author on this subject, and then went on to say that there were not more than a hundred dry-fly fishermen in the whole United States. Today, it would be unusual to meet a fly-fisherman who had not used it. It can be found anywhere in this land. In fact, it has been carried to every corner of the earth. It is no longer a novelty in America but it is certainly novel in many of its expanded or deleted parts, having acquired many peculiar forms and aspects never contemplated by its originators.

When the faint reverberations of the upheaval on the English chalk streams finally reached us around the turn of the century, the dry fly had already become established as a regular and accepted practice on those waters where it had pursued its orderly, dragless course for many years. The story of its revolutionary appearance, its birth and genesis, has been told many times and need not be repeated. It is enough to say here that it came to us in a completely formalized state, the lawful child of entomological accuracy, representing a definite attempt to copy the natural insect as human vision saw it, and used as a specific for the natural fly of the moment. When it finally arrived, a tiny bit of flotsam lapping gently at our shores, it was retrieved by a few visionaries who gave it a new and different meaning.

What a humble entry it made! But that is often the way with important things—things like the landing of our Pilgrim Fathers or the discovery of the western hemisphere by "iron men in wooden ships."

How well a man like Theodore Gordon fits this formula. Gordon the recluse, Gordon the sickly and forlorn, eyes burning bright with the fever of ill health but iron willed and glowing with the drive of the born researcher. Gordon, in his cold sickroom, muffled and cloaked against

5

the winter chill, warming his hands over a little stove so that he might be able to tie a few more of his beautiful dry flies.

Now there is a picture for a great artist to paint, some day, and it would be enshrined forever in the hearts of all true lovers of the dry fly. There are other names to conjure with too, legendary names which are part of the early American dry fly, names like Hewitt and La Branche and Gill.

Like many other cases of immigrant ideas, the wholesale adoption of concept and form is often accompanied by a period of stasis when there is neither advancement nor retreat and the adopter is largely occupied with the problem of acquiring a knowledge of fundamentals. Not for long, however, among these few enterprising and imaginative anglers who applied original thought and craftsmanship, and the results can be seen, more often than not, in the shape of the strange mutations used on all of our waters the length and breadth of the land and one or all they come under the appellation of dry fly.

It would be worthwhile and instructive to examine that dry fly which one often sees today, curling outward at the end of a leader and falling to the water in various degrees of ostentation, according to the beliefs of he who delivers it. It may ride quietly for a moment, then it may be skated or fluttered, then again it may be cast repeatedly to simulate a hatch, and very often it can be found bobbing and swooping on the surface of brawling mountain torrents. Plainly it has become a creature of many habits far removed from the conventional groove of its foreign birthplace and its lineage has become suspect, for the dry fly of today is largely a mongrel of sorts, wearing a coat of many colors under which are hidden a number of seemingly divergent views on imitation or nonimitation, as the case may be. It is clear that the use

6

of the dry fly has been extended to any and every kind of trout water, but there are distinctive changes in form, too.

NEVERSINK SKATER

Relieve the so-called standard pattern of wings, body, and tail, lengthen the hackle somewhat, and a spider is born. Add a considerable amount of hackle for the entire length of the hook shank, a spicy turn or two of contrasting color, and the bivisible emerges. Alterations to these fundamental changes have created an endless variety of patterns in different shapes, sizes (generally very large), and colors. Some bivisibles have forked tails and some do not. Some are tied with wings and some without. Small wings and a small body are often added to the spider, and it becomes a variant. Alternate bands of contrasting hackle sandwiched together become something other than a bivisible, to wit, a "fish finder." Among these must

BIVISIBLES

be included an entirely new branch of the family tree, represented by the hair-body and hair-wing flies of more recent invention. Through all of these and many more can be traced the lemon-speckled thread of the Gordon influence, a very thin thread now and somewhat wavering, but still an active ingredient in the composition of the American dry fly. It is a distinctive dry fly and derives its characteristics from its early American environment. It is the dry fly of the rough waters; it is the skating fly, the fluttery fly, the fly with which to create a "hatch of frauds." It is the fly of impressionism; the fly with which to fish the water instead of the rise, for it is the by-product of those times and places when trout are not feeding on the surface and where no surface food is present. Other times, too, where the water is rarely productive of fly life and the absence of rises is the rule and not the exception. Those who like to fish dry at all times, on such waters, need not concern themselves with an exact imitation, for there is nothing to imitate. Indeed the

VARIANT

HAIR WING AND BODY

FAN WING ROYAL COACHMAN

FISH FINDER

7

use of a purported imitation in these circumstances is a mere affectation and rivals the practice of jousting with windmills.

In these and similar cases there is no limit on the imagination or ingenuity of the angler in the matter of pattern or even presentation. He is free to exercise the greatest license and is bounded only by the limits of his own endurance and the tolerance of the trout. He may excite them with a bristly bivisible, tease them with a skating spider, cajole them with fan wings, or annoy them with rat-face McDougalls.

YOU NAME IT!

All of this is good, and suits a trout that was brought up on *The Dry Fly and Fast Water* or *Practical Dry-Fly Fishing,* but there is another kind of fly, too. It is the placid fly, a rather quiet fly, poised and serene, that obeys the stringent law of the current, no less committed to its destiny than the ancient gladiator who cried, "Hail Caesar!" It is the fly of the clear calm waters, the slow smooth currents, the long glides and flats, but most important of all it is the counterpart of the hatch, that strange phenomenon of nature which is the original sanction for the dry fly and all theories of imitation; for there are trout and there are waters which differ greatly from those which established early American practice, and the character of the stream and its fish always establishes the best pattern of fly and the best manner of fishing it, not the fisherman.

More specifically, there are the calm waters as opposed to the rough waters, the rich weedy waters as opposed to the rocky, barren waters; thin underfed trout against fat well-fed fish, and limestone versus freestone streams, all of which is sharply contrasted and plainly reflected in the attitude of the trout and their feeding habits.

These opposites cannot exist without a corresponding contrast in fishing technique and patterns of fly. Of

8

course, there are variations between these extremes which partake of a mixed character and sometimes there is a bit of overlapping in method or pattern, but generally, the established practice for one type of water does not apply to another. The placid fly as well as the active fly has its exponents, and nowhere is it more firmly entrenched than on the rich limestone waters which are probably the most productive of fly life and well-fed fish; that means a more significant relationship between the hatch and the artificial than exists on many freestone waters.

Now the problem of the hatch is the same wherever it exists, from the Little Lehigh River, famous limestoner in eastern Pennsylvania, to the Owens River in California, and its meaning to the trout is the great concern of the dry-fly angler who must deal with it.

The mechanics of the hatch and the rise are easily observed and too well known to be in doubt. In the early stages there may be a few duns and probably no surface rises, and the trout may be occupied with the drifting nymphs just prior to the main hatch; or there may be an infrequent rise in the backwaters or eddies to spinners, which had fallen the night before. When these preliminaries have ceased and the hatching duns become more  numerous on the surface, the trout may begin to show an interest in the winged insects. With a few exceptions, the natural insect is an inconspicuous thing, usually of a subdued coloration. One of them will attract little or no attention, five or ten of them hardly more, but perhaps twenty or thirty of them will arouse the trout to a consciousness of their presence, urge them beyond the initial curiosity stage, and the rise to the developing hatch begins. With some minor variations to this theme, this is the usual and correct procedure.

Objectively, it is plain that the hatch results in the

establishment of a pattern of identity, confirmed again and again by the appearance of similar insect forms passing before the trout minutes, even seconds, apart, with a given size, color, and arrangement of parts, each behaving in like manner. "John Doe, his mark" and an X appended thereto is the ancient and time-honored manner of signing legal documents in cases of illiteracy, and is aptly paraphrased by "Ephemera, his mark" with proper indentations on the surface of the water. So the trout know him.

Of course, such a thing happens only after the trout have seen enough of them, and it should be much easier to fool them at the beginning of the rise than at any later stage. Halford probably stated it correctly when he said that in the early part of the hatch the trout do not yet *have the color.* Halford used the italics to underscore his meaning, for he was not referring to color itself but to the fact that the feeding fish had not established in their minds the identity or recognition of a palatable insect which they willingly accept without deviation at a later stage. Repeated experiences with the same species of insect during the progress of the hatch cures this defect.

It is nothing more or less than the result of familiar stimuli repeatedly traveling along the same neural paths and creating a ready response from the reflexes of recognition, something like acquiring a habit. It is not a very great intellectual accomplishment, but it has its advantages. It certainly makes it more difficult for a trout to accept the odd things offered to him by fishermen during the hatch, and because of the inflexible nature of his mind, unencumbered by imagination and narrowly bound by the realism of a harsh society, it causes him to react undeviatingly to food in the form of insects of a recognizable, safe, and acceptable standard; and there you have the fundamental reason for a "strict imitation."

10

The hatch pleases a trout in other respects too, since it represents food in such quantity and quality as to suit the economy of his existence. During the whole of his life he is forced to practice extreme economy of movement, never exerting himself unduly. His frugality in this respect makes even human indolence seem like sheer extravagance. Any number of mammals—a small dog, for example—of the same comparable weight and size are capable of expending a tremendous amount of energy continuously, for hours, even days on end. Ordinarily, a trout can be played to complete exhaustion in a matter of minutes with the gentlest resistance, even when he is in the best of condition.

Saving and gaining energy is always a problem with him, particularly in running water, where he must constantly search for a suitable resting place, one that requires the least possible effort to maintain his position in the currents. There are other demands too, on his pitifully small store of vitality. Barometric pressures, temperatures, and chemical changes in the water of an adverse nature constantly plague him, forcing him to make internal adjustments of such a critical nature that his life often hangs by a thread. His is a double-entry way of life and he is eternally threatened with bankruptcy.

It follows, then, that it is a prime necessity for him to obtain a maximum amount of quality food with a minimum expenditure of energy; otherwise his existence would be a biological error, a predicament against which he instinctively rebels. In this respect the hatch pleases him very much, for it means that his food is being carried to him (in running water) instead of forcing him to seek it; his feeding habits at this time are characterized by the greatest economy of movement. He will choose the best lines of drift, where the concentration of drifting insects is greatest, in order to operate in the smallest

11

possible area. The hatch pleases him further by permitting him to employ a certain amount of rhythm in his feeding. Confront him with something different from what he has been eating and he is immediately faced with the task of inspection, then acceptance or rejection of a doubtful article—a process which is liable to lead to rejection, hence a break in his rhythm and a waste of energy. That, too, is a part of his economy and another good reason for strict imitation.

All of this is much more apparent in calm, clear waters, rich in fish and fly life, than in fast, broken, less fertile streams. If one fishes the latter very much or exclusively, it is easy enough to acquire a belief that any fly will do at any time and that there need not be a hatch to raise your trout; but that is because such trout are hungry and lank, and are likely to strain every nerve and fibre to get something that looks alive and tempting. Freestone trout are often so lean and empty that sometimes their stomachs contain the most astonishing assortment of indigestibles—even sticks and stones.

Such a thing is very rare in a native, stream-bred limestone trout; and what's more, he is not the kind to be teased to the surface with a simulated hatch. If you try such a thing on him, he will probably cock a jaundiced eye and grin his derision, for he is a worldly citizen who has usually dined very well on every course, from antipasto to dessert. He has never read La Branche or Gill and would not understand them, and that means that dry-fly fishing over such trout is the poorest in the world without the hatch.

Of course, there are always the exceptions to a general rule, as in the case of freshly stocked hatchery trout or hand-fed fish that have not learned to feed in a natural environment even though surrounded by an abundance of natural foods; but generally it is unprofitable and de-

12

cidedly harmful to cast the dry fly on fertile water containing well-fed fish that have not come up to a hatch. It is far, far better to fish the wet fly, and for those who like to use it there are tremendous rewards for the skillful performer. Your genuine limestoner has discovered this, with the curious result that there are some first-rate wet-fly fishermen on these first-rate dry-fly streams; but that is another story that would make a whole book by itself.

Our concern here is mainly for the dry-fly angler who is so often confronted with the problems of the hatch and the nature of its appeal to the trout. There are few fly fishermen with any experience who cannot admit failure, and the pattern of this failure is usually the same. There is an old familiar scene that haunts every fisherman who has encountered the hatch and all of its perplexities. It always involves an angler standing at streamside who sees with mounting excitement the emergence of a fine hatch of duns, and more wonderful yet, the trout splashing and feeding in every quarter! What a mixture of emotions possesses him at this time! Some anxiety, perhaps, a little confidence, no doubt, and a half-formed plan of attack involving approach, position, casting, and, lastly, the fly pattern. At this last point, grave doubts and questions began to assail him. In all likelihood, after a hurried inspection of the contents of his fly box, he will select a reasonable facsimile of the natural fly and, if it  fails to attract the fish on the first few casts, exchange it quickly for a different offering, selected with a little more intensity than the first one. If again failure occurs, he is likely to resort in some desperation to a favorite pattern, one which afforded him some success on a former occasion, abandoning entirely any further attempt at specific imitation. If this sturdy favorite proves to be no better than the others, his confusion is complete and he proceeds to offer any and all of the wildest creations of the

13

fly-dresser's art, including his own. At this stage, goaded to desperation by his mounting number of failures, he is overtaken by the fatal realization that time is short and opportunity escaping him. His distress accompanied by much fumbling, poorly tied knots, and faulty casting finally becomes a state of "fine frenzy." Our unhappy angler, veteran though he may be, all but throws the fly box at the trout. His agony of effort is terminated only upon cessation of the hatch and the rise.

No one is immune to this experience and those who have suffered it can, no doubt, recall the blank period that usually follows, when the angler stands listless, staring and seeing nothing, possessed by a detached and bemused air, filled with the futility of his experience. When he finally turns away, slowly and aimlessly, a gradual consciousness of his position invades his mind, and if he is the emotional kind he will probably berate himself, his ineptness, and the perversity of fish nature, and finally swear that he will give up fishing and play golf!

How often this same scene is repeated in the life of a dry-fly fisherman, this familiar bit of streamside drama which might well be entitled the Serio-Comedy of the Hatches.

It is not always easy to say why failure occurred, but it happens often to the man who knows how to place his fly accurately, on the proper line of drift, with the finest gut, and without drag—often enough so that we may fasten our suspicion upon the pattern alone.

Now you may argue successfully for a "strict imitation," but obtaining it is another matter. The rejectionists would have had the better of the dispute had they based their argument solely on the difficulty of accomplishing such a thing rather than upon denying its necessity. Worse, far worse is the problem of defining the term, for

14

if he is aware of all the implications, he is a brave man who dares to attempt such a definition. It cannot be done in the dictionary way; above all, it cannot be done in the way of Halford, Ronalds, and others of ancient fame, for they spoke of imitation in terms of human vision and comprehension, supported only by the prop of entomology. That way alone lies grave error, since it does not take into account the vision of the trout and the geometry of the underwater world; and the study of entomology stops short, far short, of the approaches to these considerations, which are the dominant factors in devising imitations.

It is better not to define the term at all, since no one can be sure that he knows all of the elements involved. It is better to speak of imitation in terms of the presently discovered elements, and to improve it as others are added; but first I think that it is proper to speak of the kind of fishing that needs a "strict imitation," an imitation which is intended to be a "deceiver" and ought to occupy an equally important place with the "attractors" in the fly box of every fly-fisherman who is prepared to deliver a well-rounded appeal to the trout, wherever it can be found.

Chapter 2

FISHING THE DRY FLY ON QUIET
WATERS: A Character Study

MINERVA sprang full grown and armed from the brain of Jupiter. The mythical goddess did not enjoy a more complete birth than many of the limestone streams of central Pennsylvania. Issuing with great force and volume from the Stygian caverns in the soft lime rock, they present an aspect of maturity no less entire at the immediate source than at any other point in their course. Constancy of flow is their great virtue, implemented and made secure by union, over- and underground, with other streams of a similar speculative origin. Blessed by nature with a boundless store of the elements which are conducive to the flourishing of underwater life, they are unsurpassed in their capacity to nurture and maintain a

17

stock of trout and other fishes in extraordinary quantity and growth. Weeds are profuse and varied, insuring the continuance of those lesser underwater creatures which thrive in their midst and constitute the finest larder ever designed by man or nature for the trout family.

There was a time, according to the ancient annals, when the average weight of brook trout caught in these waters was near two pounds and the yearly total of such trout for a single fisherman was numbered in the thousands, an impossible thing now, in these sterile times of premature harvest and inordinate demand by an ever-increasing horde of fishermen. Yet the inherent richness of these waters remains unchanged, sometimes evidenced by the astonishing size and condition of brown trout, recently introduced, which have escaped for a time the searching and ever present barbs of the relentless multitude.

There was a time, too, when the encroachments of civilization and the urgent needs of man did not interfere with the normal purity and clarity of these streams, as they have of late, with the regrettable result that miles of these valuable streams remain unproductive and barren of fish life. Wherever they can be found in their original state of excellence they continue to be the favorite haunts of fly-fishermen as they were a hundred years ago and more, particularly the dry-fly fishermen, who have discovered them to be pre-eminently suited to the use of the dry fly by virtue of the habitual presence of surface insects and the willingness of the trout to rise freely to them.

Fortunate indeed are those who, by force of circumstances or choice, are able to pursue and enjoy a form of angling which is exactly suited to their inclinations. When a happy condition of this sort exists, it seems almost inevitable that the angling practices of that locality

18

will occasion a constant and intelligent study by its devotees in order to create the refinements which so greatly increase the pleasures of angling. In some instances there may be a sufficient number of those individuals, keenly observant and studious by nature, who bend their combined efforts in study and experiment to evolve and formulate a new and interesting method of angling.

By all accounts, these are the factors which gave rise to the birth and bloom of the art of dry-fly fishing. Cradled and nourished in the chalky downs of the Hampshire district of England by men whose names have become a byword to fly-fishermen everywhere, it has risen to a plane so lofty as to border on the aesthetic!

It seems quite clear, in retrospect, that the dry fly would never have occurred in its present exacting form without the existence of the chalk and limestone streams of England or streams of a similar nature elsewhere. Unquestionably this was the stage, the proper setting arranged and prepared for the advent of men like Halford, Lord Grey, and Marryat. Without this background—the slow-flowing, smooth-surfaced, limpid waters of Test and Itchen—without the luxuriant water weed and the consequent abundance of water-bred insect life, notably the Ephemeridae, there would have been no plastic, no clay so to speak, for these men to shape and mold. There would have been no dry fly. It does not matter that these men might have come in an earlier or later century. The important thing was, and is, the existence of these unique rivers.

So, too, in the present instance it is appropriate to establish a proper basis or background to authenticate, as it were, the origin and development of new dry-fly patterns and their application to these and similar waters. In more recent times these efforts have been centered

19

largely on the beautiful Letort, at Carlisle, Pennsylvania, and the sister streams nearby.

What shall I say of the Letort? Certainly it is beautiful, not with the wild beauty of our mountain freestone streams, decked out in their garish display of laurel and rhododendron, but rather with the calm and serene beauty of pastoral scenery. Nestled in a little valley, with gently rolling hills on both sides, it meanders slowly and evenly, its placid surface hardly ever ruffled even by the westerly winds which prevail thereabouts. Even a heavy downpour creates no severe change other than a barely perceptible fullness and a slight milkiness which disappear in a short time, a matter of hours.

I never cease to marvel at this phenomenon, particularly when I recall a certain fishing hut, which has its foundations resting on a level bank about ten feet from the stream and hardly six inches above the level of the water! Yet, within my memory and that of many who often fish there, no water has ever been seen underneath the hut.

The reader must not suppose, however, that these features indicate a lack of character in the Letort. It could not be so, in consideration of its great depth and vast weed beds and channels exerting their subtle influence underneath the calm exterior, creating currents and crosscurrents, intermingling them in a sinuous manner which, barely noticeable to the casual visitor, causes no end of astonishment and despair in his efforts to overcome drag. The Letort is a hard taskmaster and does not treat lightly any violation of dry-fly technique. Any suggestion of drag, heaviness in the cast, or thick gut is magnified many times on such a calm surface, and the penalty is absolute and total failure.

On the other hand, a successful angler is rewarded handsomely in the event of a proper approach, position,

Fox's meadow in the Letort, as it looked before "progress" came to this valley. A part of the old fishing hut is shown at the left. A huge four-lane highway bridge now spans the area where the fisherman stands, and the fishing hut no longer exists.

Otto's meadow—still a very productive stretch of water. An eight-and-one-half-pound native brown trout was caught on a fly in the left foreground area.

and cast. It could not be otherwise in a stream where trout grow to such noble proportions in a short time and rise freely to the smallest of dry flies, and where the rises themselves are of such a character as to startle the most phlegmatic of natures. I must confess that I have never become used to them, even the quiet and dainty rises. One moment the artificial is floating quietly and gently, and the next instant there may be a sudden heaving and bulging of the water, accompanied by a sound of basso profundo depth that cuts into the consciousness with a sharpness which is a little unnerving, but also exhilarating. Even where the rise form is a tiny sip or dimple, awareness comes suddenly to the angler from the violent lunge of the fish and the ratchety sound of the reel. I suppose it is the severe contrast of the quiet of the stream and its surroundings with the commotion and noise of the rise which affects me in this manner, for I do not experience the same thing when fishing rough water. Then, too, there are those occasions, which are very frequent on the Letort, when I can observe every movement of a feeding trout—see him aquiver the instant my artificial lands a foot or two in front of him, watch him detach himself from his observation post, undulating backward and lifting slowly to take the fly at the precise instant they both meet at the surface, as though the trout had calculated the interception by triangulation! These are the worst times for me. Being somewhat nervous by temperament, I can not subdue the building up of nervous pressure engendered by the visible and deliberate rise. The usual result in these cases is a violent reaction on my part intended to be a strike, something I fervently wish I had never learned, and the matter is concluded by a sudden parting of the ways between the trout and me. Again there are those moments of indecision on the part of the trout when the backward undulating movement is

21

continued for perhaps three or four feet from the usual point of interception, the nose of the trout barely touching my artificial while he scrutinizes it carefully, and all the while my anxiety is mounting with the realization that the point where free float ends and drag begins is fast approaching. In other respects the Letort presents an aspect of constancy in hatches and rising fish which is extremely gratifying to the angler-entomologist-re-searcher.

On some waters the prospects of finding trout regularly and habitually taking surface food are slim indeed. As for daily routine, it simply does not exist, being the exception and not the rule. Anglers who fish those waters and who like to employ the dry fly are constrained to fish to a position rather than to an individual fish, as I understand it, searching, with a fly of impressionistic design, each pocket and eddy, or any likely place where a trout might be lying in wait to seize a chance morsel as it goes by.

On the other hand, trout of the limestone waters constantly exhibit a preference for surface food that is ritual-like in point of time and place. In recent years it has been an object of special interest to me to determine the extent of this habit. I have taken every opportunity during the course of my fishing excursions to note the time of day and the number of successive days on which rising fish could be observed, and I lost no opportunities to make similar observations on those vacation periods when I lived on the stream itself for as long as a week at one time. On these latter periods it was my custom on leaving the hut in the morning to take readings of water and air temperature, note barometric pressure, and then patrol a beat of one-half mile of water, taking note of the time and place of any rises during my tour; I re-

22

peated this whole routine once in every hour during the day.

Water temperature seemed to be the most decisive factor in determining the regularity of rises. The temperature at seven or eight o'clock in the morning always gave a reading of 50° or 52° F. Each successive hourly reading thereafter revealed a rise in temperature of 2° F., until about two o'clock in the afternoon when the peak was reached at 60° or 64° F. At approximately eleven o'clock, when the readings were 56° to 58°, the first rises could be seen, few in number and taking place at longish intervals. As the temperature increased the number of rises increased, the intervals between rises of the individual fish shortening considerably; the peak of activity was reached when the highest reading of 64° F. was taken. It was interesting to note that brown trout were not the first to begin the search for food. At the lower end of the beat there is a part of the stream which seems to be especially suited to brook trout, for a fair number of them can always be found there. The inception of the feeding period at ten-thirty or eleven o'clock, concurring with the temperature reading of 56° or 58° always took place at this location and was evidenced by the movements of brook trout only. At one or two o'clock, when the temperature of the water gave a reading of 62° to 64°, the brown trout made their appearance and from that time until approximately five o'clock both species fed with sustained regularity.

This performance continued day after day; as nearly as I could determine, the only factors that caused any interruption in these conditions were changes in barometric pressure, and thunderstorms. These fish are positively allergic to thunder and other heavy vibrations of any sort. An occasional blasting operation in a nearby limestone quarry will put them down for such a long time that the angler is forced to occupy himself else-where until their confidence is restored and their feeding resumed.

Armed with this sort of knowledge, those who fish these waters are in an excellent position to adjust the time and place of a fishing jaunt to conform to the well-established habits of these trout, but I hasten to advise that there is no guarantee that they can be caught!

Familiarity with all of those facts has provided me with an unending source of amusement at the expense of a stranger or infrequent fishing guest. It is no great trick to inform him in a sober and offhand manner that he will find trout rising in a certain place at a certain time and confound him further by sending him off to try for a fish that rises at a certain hour just three feet below a box elder and two feet from the right-hand bank. I fear that these doings have earned for me, in a few cases, a repu-tation for clairvoyance and wizardry in trouting matters that fits me badly, particularly since every once in a while I am invited to try for the fish and fail miserably to take it!

All of this makes this sort of fishing a reasonably con-stant and predictable affair not only with respect to the activity of the trout and the state of the water but also with respect to the presence of insect hatches, water-bred, and insect flights, land-bred.

Let others less fortunate envy the British and their olive dun and its spinner, that wonderful fly which by common acknowledgment seems to thrive and appear in great numbers on all the English dry-fly rivers. The olive is very scarce in America and its imitation is of question-able value, but in these parts dry-fly anglers may rejoice in the existence of a pale, watery class of mayfly, both the pale-winged and the blue-winged variety, which forms the backbone of dry-fly fishing on limestone

24

waters. In a very good year, when hatches are especially heavy during the peak of emergence, they can be seen streaming upward from the surface of the water, their bodies catching and reflecting the full rays of the sun, winking and glowing like a thousand hot golden sparks. There is cause to rejoice further in the fact that they emerge almost daily from early May until the end of the season. Oftentimes, their behavior on the surface of the water is characterized by considerable effort and exertion in order to dispel the nymphal shuck until finally, they glide along in full repose with wings erect, like little ships with sails unfurled. That all of this is a form of conduct highly absorbing to fish as well as to anglers is confirmed by the presence of rising trout everywhere, which take the insect with a peculiar noise, much like the sound of a cork popping out of a wine bottle.

Equally interesting to both angler and trout is the appearance of the Hendrickson, locally called Slate Drake, which usually obliges the grateful angler by emerging on or about that auspicious day, April 15, the beginning of the trout season. The pale wateries might be depicted as dainty, the Green Drake as a clumsy, lumbering beast, but gallant is the word which describes the Hendrickson.

No sight is prettier than to see a string of these creatures come sailing along, wings erect and proud and slanted at a rakish angle, bodies rocking mildly with each little wavelet, causing the tall wings to tip and bow, now this way, now that, as they turn with the current.

On the Yellow Breeches Creek, before the first Hendrickson is seen the caddis flies have already put in their appearance. The variety and number of caddis at this time of year is astounding. Light-colored ones and dark ones, small and large, they fill the air above the water in countless hordes, dipping, fluttering, and flying aimlessly

25

as caddis flies will. They are often so thick that the form of an angler a hundred feet away is reduced to an indistinct blur. It is not unusual for an angler to leave the stream after a day's fishing at this time and discover that his waders, from top to toe, are covered with a solid mass of green egg sacs, acquired from the deposit of the industrious caddis. The memory of such an experience is indelibly inscribed on the mind of the unlucky angler when he discovers the tenacious character of these egg sacs in his futile attempt to remove them.

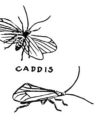

CADDIS

These various caddis represent a very powerful attraction to the trout, causing them to gorge and regorge as they see fit. But let the first Hendricksons make their appearance and the order of the day is changed completely. The caddis are entirely abandoned and the trout immediately direct their attentions to the Hendrickson, displaying a ravenous appetite, which is astonishing in view of their previous gluttony.

Here indeed in an extraordinary situation. The great profusion of caddis, the advent of the Hendrickson, the change of taste by the trout—all of these are conditions calculated to produce doubt and uncertainty in the mind of the angler. An artificial which easily deceived the trout taking caddis earlier in the day is no longer successful; and if, as the angler suspects, the trout are taking Hendrickson, how is he to know when and where? Fortunately, some order out of this confusion is possible. If the angler carefully observes the rise-form of individual trout, he may learn that there is a remarkable difference between the rise to caddis and the rise to Hendrickson. Trout taking caddis do so with a slashing strike, throwing a bit of spray accompanied by a thin, watery sound; in the case of the Hendrickson there is a deliberate sucking rise, causing a single boil or ring on the surface and creating a sound which has a deep, resonant quality.

26

But a word of caution is necessary here, for oftentimes a trout which has just diverted his attention from caddis to Hendrickson will continue to exhibit a rise-form of the slashing kind for a short time. Once the feeding on Hendrickson is fairly under way, the trout will gradually re-form his pace until it resembles the correct rise-form to the Hendrickson.

It is a regrettable day, a regrettable hour when the last of these fine insects is seen at the end of the emergence period, which lasts for perhaps three or four weeks. Given a good head of trout in the stream and provided that weather and water conditions are suitable for the entire period, there is no fishing to any insect which is more likely to stir the interest and imagination of the angler.

Do not suppose, however, that the passing of the Hendrickson and the coming and going of the pale wateries put an end to the anticipation and fulfillment of the dry-fly angler's chosen pursuit. Even in late season, when the larger mayflies are less and less in evidence, the surface of these waters is literally carpeted with terrestrials of one form or another, providing a source and inducement for surface feeding that rivals the earlier duns and spinners; and in some ways it must be admitted they are even more important to the dry-fly fisherman than the Ephemeridae because of the attraction of their larger, meatier bodies, analogous to the same attraction which exists for the trout during the season of the giant mayfly, Ephemera Guttulata, or more familiarly the Green Drake, which occurs in tremendous quantities on our limestone waters. In particular, I would commend to anglers of the limestone waters everywhere the marvelous fishing to be had during the grasshopper season, a phase of dry-fly fishing little known but highly developed in these parts, which makes the short baccha-

GREEN DRAKE

nalia of the Green Drake very pale by comparison, if only because of the lengthy period during which grasshoppers are available. In like manner, I can speak of the newer and equally fascinating form of dry-fly fishing provided by that otherwise unwelcome newcomer the Japanese beetle, which at present is largely confined to the Eastern seaboard but, by all accounts of the entomologists, is gradually but inevitably moving westward and in time should appear on all the waters of this continent. True, it spells agricultural calamity wherever it 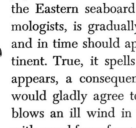 appears, a consequence deplored by this author, who would gladly agree to its complete extinction; but if it blows an ill wind in the agricultural quarters, it blows with equal force for good in the angler's favor. During the annual period of its presence, lasting for perhaps three weeks, it appears in such tremendous quantities as to belie description, clinging to and devouring avidly every form of leafy growth. In some cases, where its attentions are concentrated on an especially palatable object such as a buttonwood tree, the ultimate result, in a comparatively short time, is similar to the midwinter starkness of hardwoods, but covered with lacework. The activity of these beetles is not confined to inland shrubbery but progresses to the very edge of all waters, where  they can be found on the tips of any overhanging of trees or shrubs which inhabit the streamside; and in consequence, they are constantly dropping and being shaken onto the water by various influences, creating in this fashion a situation much like a hatch of duns, with the same succulent appeal to the attentions of the trout. Do not suppose that the presence of beetles and the activity of trout are restricted to the edges of the stream or near the banks. Let there be a current angling away from the bank or an obstruction causing such a current, and a stream of beetles is forthwith diverted even to the

28

center and farther on some of our widest rivers, stirring and exciting a chain reaction of feeding activity. Then let there be other currents at these farther points, splitting and splitting again, diverging fanwise, carrying beetles in every direction to every feeding position in the river.  These are not my observations alone but also those of many anglers of my acquaintance who wade and fish the broad Susquehanna for bass—reliable fellows, who can confirm these statements and who have done so by making numerous autopsies on bass caught in mid-river, disclosing stomachs gorged and distended by Japanese beetles and nothing else.

Many of those who live outside the orbit of the Japanese beetle may experience little or no interest in this account, and naturally so when there is no opportunity to observe it and fish to it, but I dare say that, if and when it finally arrives in consequence of its steady migration in those localities presently uninfested, there will be ample occasion to recall these words with a new significance.

Equally important, and no less abundant on limestone waters—perhaps more so than grasshoppers and Japanese beetles—is another class of insects so tiny in form as to be almost invisible on ordinary inspection, giving rise to the existence of a curious phenomenon that involves some of the most fundamental concepts of fish habits and angling technique.

Most of us, as a rule, regard 18 and 20 sizes in dry flies as being extremely small, but this is a situation that demands even smaller artificials. In fact, size 22 is still too large. But besides the matter of size, there is a strange twist in the manner with which they are taken, something which is seldom observed or accurately appraised.  But once a friend and a very keen angler offered this remarkable opinion: that a great many times when we

29

think a fish is nymphing, it is really taking very tiny insects, living or spent, from the surface. A statement of this sort, casually given, is likely to be accepted as a rare happenstance or an unusual incident which an angler might encounter at one time or another, but which is not really worthy of the angler's attention. Whatever may be its application on other waters, however, for the quiet meadow streams it is a monumental pronouncement representing and embracing a tremendous segment of regular angling procedure.

To explain the matter fully, I must revert to my early days on the limestone waters, when my visits were rather infrequent and I had no more serious purpose than to catch fish. I can recall now with some amusement the perplexity of myself and my friends on those days when we were confronted with feeding trout everywhere in sight, but trout feeding in such a manner as to render indeterminable the kind and source of food and the level from which it was being taken. At that time I did not have a particularly good knowledge of the various rise-forms, but believed from what I had read and heard that the trout were engaged in nymphing. The rise-form in evidence seemed to fit the standard description exactly —a slight humping or bulging of the water, no apparent  break in surface, and no surface food visible. I flattered myself at the time for having solved the problem so neatly and immediately proceeded to offer to the fish every conceivable form of artificial nymph that I could find in my box, including some that had no counterpart in nature. All to no avail. This same comedy was repeated many times and my disappointment increased with every occurrence, except for a short-lived interval of renewed hope when a new pattern of nymph came on the market and I tried it, only to fail again.

At one time I believed that trout were feeding on

snails in this manner, for they can sometimes be seen floating along just beneath the surface of the water; * and because I was aware of the great quantity of these creatures in those waters and the trout's fondness for them, I had a right to assume that they were the source of food and the reason for the peculiar rise-form. But subsequent observations indicated that the bulk of this food is plucked forcibly by the trout from stones and abutments to which these animals cling, while only a small proportion of them are taken when floating free. At Big Springs and in the Yellow Breeches, certain gravelly and rocky stretches are so densely populated with snails that the wading angler often experiences a distinct crunching sensation underfoot, but they are dislodged from their anchorage so seldom and in such small numbers that they could not be the cause of the mysterious surface activity of the trout.

Matters rested this way for some time until one memorable day a few years ago, when I was again confronted with this problem. I was on the Letort at the time in the company of Charles K. Fox, whose unusual skill and resourcefulness were severely challenged by this peculiar problem that I had failed to solve. It was shortly after he had acquired the beautiful stretch of the Letort which he now owns, and since he was anxious to familiarize himself as much as possible with the sporting possibilities, we spent a considerable number of days together, making many observations, fishing continuously and enjoying ourselves tremendously. Uppermost in our minds was the age-old problem of the bulging, humping trout.

* It is my impression that snails do not float freely in the water. Upon examination, these floating objects were discovered to be empty shells which were probably cast off or, abandoned by the snails, leaving them light and buoyant enough to rise to the surface and float away.

On this day they were performing in the same mysterious fashion and we were making the usual offerings—nymphs of all kinds and sizes, wet flies whole, and wet flies trimmed—all of them fished with change of pace, dead drift, jerked in short strokes and long strokes, fished high and fished deep, but none of it produced a single fish. There were at least a dozen good trout within my immediate vicinity and I went from one to another in the same futile routine. Charlie was engaged in like manner, passing me at intervals and returning to try for another fish and giving me a questioning glance each time to which I am sure my own hopeless, unspoken query was sufficient answer.

The day might have ended in the usual manner except for the fact that my disillusionment reached a critical point and I quit fishing entirely, content to lay aside my tackle. Stretching myself prone on the bank with my eyes not more than one foot from the water, I beguiled the time by watching a small trout nearby feeding in the manner of his brothers. I had no particular object in watching the water, but I do it often, enjoying the mesmeric effect of smoothly gliding water and the pleasant lassitude it induces. I do not know how long I stayed this way on the bank, but it was long enough for me to acquire a sense of unreality about my surroundings and the water before me so that eventually I began to see things on the water which should have been only a figment of imagination—tiny little duns, dozens of them, some struggling to dispel the nymphal shuck, others sailing along in perfect repose with wings unfurled; other creatures too, minute beetle forms, moundlike in shape and shiny in appearance. I allowed this picture to dwell in the half-consciousness for some time until a sudden sharp awareness invaded my mind, causing me to focus my attention on the water more closely. Gradually a suspicion

32

of the truth dawned upon me. The sight before me was not figment, it was real.

A mild excitement stirred me and I hurried back to the fishing hut where I found the thing I needed so badly—a square mesh bag, with ⅛-inch mesh, which seemed a little too large, but doubled made a mesh of approximately ¹⁄₁₆ inch. I found two sticks of suitable length and thickness which I inserted in the bag, and by holding the two sticks apart, I was able to maintain sufficient tension on the cloth to provide me with a satisfactory screening device. With this equipment, crudely fashioned but seemingly practical, I returned to the stream and chose a likely place for my experiment, a place where the flow of water was concentrated in a narrow channel, in order to intercept the greatest possible amount of surface drift. I inserted the net about half its length below the surface and was considerably annoyed to discover that so fine a mesh could offer so much resistance to even the gentle current of the Letort; the resistance caused the water to roll back upstream for a short distance, where the downstream current caught it and forced it to curl around either side of the net, carrying with it all of the flotsam I had hoped to entrap. Eventually I discovered that allowing the top of the net to lean downstream at a considerable angle lessened the resistance a great deal, thereby permitting a goodly amount of surface water to seep through the mesh and deposit the mysterious burden which interested me so much. I maintained this position for perhaps fifteen or twenty minutes, enduring the discomfort of icy water on my hands and wrists, then withdrew my catch and, carrying it to a place where the light was good, proceeded to examine the deposit as carefully as possible.

It did not take me long to find that my suspicions were confirmed and I was properly elated. Clinging to the fine mesh of the cloth in great numbers and arranged in a

33

straight line across the netting where it had coincided with the surface current were many tiny insect forms, unbelievably small—perfectly developed little duns no more than ⅛ inch in any dimension; beetle forms, no more than ³⁄₃₂ inch long; reddish-gold ants in the winged state, so small that their slender waists were almost invisible; but most amazing of all, a respectable number of large black ants, fully ½ or ⅝ inch long. I could not understand immediately why I had never detected the presence of the large ants in my previous examinations and I returned to the stream again to discover the reason for this defection. After a careful inspection, I found to my surprise that the large ants did not float on the surface of the water but drifted awash, partly submerged and flush with the surface, wings sodden and blending with the background in such a way as to make them completely invisible to the angler.

The significance of these revelations was plain to me, and I envisaged a whole new line of thought and approach; but the task at hand was not completely finished, for it was imperative that I should establish the connection between the mysterious rises and the newly found food forms. I called to Charlie, who was engaged elsewhere at the time, acquainted him hurriedly with all that had occurred, and explained the necessity of obtaining a few trout for the purpose of making autopsies. He readily agreed, and we renewed our attempts to lure and capture these trout with livelier hopes, but with a different approach, although in a somewhat apprehensive state of mind because of the rapidly failing light. I searched in my fly box for a very small fly, dressed on the lightest of hooks, and finally selected one in size 20 with a reddish-tan fur body and a rusty dun hackle. I trimmed away all of the wing and almost all of the hackle except for two or three barbules on each side of the body, attaining in this manner the desired emphasis on the body and insuring the likelihood that it would float low and flush with the surface in the manner of the naturals. I lost no time in presenting this unusual lure to a feeding fish, noting quickly that it floated properly, for I could barely see it on the surface of the water, and promptly raised, hooked, and landed a trout of medium size.

In like manner, I succeeded in taking three more fish and lost a fifth in the weeds before I could turn its head and skulldrag it across the weed beds as I had done with the others. It was not pretty fishing, and ordinarily I  would not have been guilty of such forceful and ungraceful manners on a trout stream (and I confess, too, that my trout were of a size which should have been returned to the water for the good of the fishing); but I was desperately in need of these specimens. The autopsies were performed without delay, before the digestive processes had time to render the stomach contents unrecognizable, and I found to my satisfaction that the entire bulk of the food taken consisted of the minute forms which I had netted earlier. I might add that the very smallest of the ants were present in great numbers, with a good representation of the wee duns and only a few of the very large ants. I cannot say for certain that any one of these insects was preferred over another by the trout, who seemingly took whatever came to them in the line of drift. The predominance of the little red ant in the autopsies may only mean that they were greater in number than the others on the water, in the same proportion as they were ingested by the trout.

If there is any doubt about choice of the correct pattern to be presented, it should be resolved in favor of the one which imitates most closely the insect which is most in evidence at any given time, and this seems to be a sound and logical rule to follow. However, I would like

35

to caution my readers that this rule is not infallible and sometimes does not apply when there is a variety of large insects on the water. There are some seasons, notably the Green Drake season, when the trout often prefer a smaller, darker dun emerging at the same time; and I can recall a similar instance, which takes place regularly on the Yellow Breeches stream in Cumberland County, when the concurring prevalence of the Hendrickson and several species of caddis flies, all in great profusion, makes it extremely difficult to determine the preference of the trout. As noted before, it is absolutely necessary for the angler to recognize the rise-form which identifies the taking of any particular species if he is to select the proper artificial.

I cannot emphasize too strongly the findings which I have reported in the preceding account of the peculiar feeding habits of trout when minute insects are being taken. It accounts for a great deal of the fishing to be had in the latter half of the season on rich meadow waters, to an extent and of a nature not heretofore appreciated by the anglers of these parts. It was an especially lucky stroke that I happened to be present when there was such a great number and such a variety of these minute insects. I have not seen an assortment of this kind since that day. In all likelihood there was a fortuitous overlapping of hatching or flight periods of these tiny creatures on that particular day. But rest assured that they are nearly always present on the surface of the stream, even though they are almost invisible. In the normal course of things they occur, a single species or perhaps two at one time, during their appointed season. Some are of short duration, although one amazing little red ant, in the winged state, appears almost daily and continues until November. The imitation of this insect in the floating pattern looms with increasing importance in fly-fishing practice on these waters; it should be tied as small as possible—size 20 short shank or 22 regular shank, or even size 24.

The little red ant still remains somewhat of a mystery; I have never actually seen it in flight, nor has anyone else as far as I know, although I have heard rumors to the effect that anglers have seen clouds of little red "bugs" on the Letort at the ungodly hour of five o'clock in the morning.

This abundance of minute forms is not confined to the Letort or the Newville water. Once when I was fishing the Yellow Breeches in May, I remarked about the appearance of the water, which seemed to be covered with a layer of fine particles, as though a giant hand had shaken salt and pepper on the surface, and I carelessly ascribed it to a shower of sooty grit and dust from the passing engines of a rail line immediately adjacent to the stream. Later in the day, I took the opportunity of examining a particularly fine specimen of the Light Cahill floating on the water, ladling it out with the palm of my hand, and was startled to discover that, along with Ithaca, my hand was entirely covered with the tiniest of the Ephemeridae. I repeated the operation several times, each time with the same result.

I can recall other occasions, on other streams—Honey Creek in Mifflin County and, for another, that wonderful dry-fly stream, Spruce Creek in Huntingdon County—where I was badly beaten by trout that performed in the strange fashion which I have already described.

Lest there should be some misconception about this phenomenon, particularly if someone should liken it to the kind of fishing which accrues from the heavy fall of spinners or imagos of the larger Ephemeridae, be advised that it bears no resemblance whatsoever to that kind of fishing. The rise-form of trout feeding on spent

37

imagos may be similar, but there is the tremendous difference that a heavy hatch of duns preceding the fall of spinners, and easily observed, forewarns the angler of the kind of fishing to be expected in a day or so. He need not be present to see and capitalize on the return of spinners, for oftentimes trout will feed actively on the following day when there are no insects in sight and when the spent forms are lying adrift in the backwaters and eddies, flush with the surface and almost invisible. If trout are observed feeding quietly in the bulging, humping manner, the angler, being forewarned, may  safely conclude that the spent artificial is the proper choice for presentation. I insist, too, that no comparison should be made with the case of trout feeding delicately on smuts, for small as they are, they can usually be seen in clouds, milling about above the water and rising and dropping in unison. If a trout waiting below and just beneath the surface can be seen making a quiet and dainty rise, it is fair to assume that he has taken one of these tiny insects and will probably take more. The circumstances are too easily related in this case to leave the source of food in doubt.

 The imitation of all of these minute insects, as well as the larger mayfly duns and spinners, has been an object of great concern to me. I feel that there is a great need for specialization in the field of terrestrial insects and the minute forms—a specific treatment, heretofore neglected, with the object of creating satisfactory artificials intended to be used when their employment is indicated, and furthermore a formulation of the proper tactics to insure their success when presented to the fish.

Some progress has been made in this direction, warranting the belief that a solid foundation has been laid, lacking only those small refinements to achieve perfection. On this score I have not hesitated to capitalize on the noble assistance of many interested and sympathetic friends. Once, when I was wandering about the Letort during the grasshopper season, I had a chance meeting with Gene Craighead and his brother Charlie, both fine anglers and both of them highly absorbed, as I was, in the spectacle of several large trout feeding with wild abandon on grasshoppers, which not only could be seen all around the meadow in great quantity but also were finding their way into the stream in surprising numbers. Our conversation was largely limited to the problem of devising a proper imitation of the grasshopper, and I confessed that up to that point I had not attained anything of a satisfactory nature. Thereupon Charlie informed me that he had considered the problem somewhat and the night before had completed the tying of several patterns which might be successful. I was inclined to be skeptical but nevertheless accepted two of them, which he graciously offered to me from his fly box. One was of a greenish-yellow coloration, about an inch and a quarter long, with brownish wings tied flat. There is a wide range of coloration in grasshoppers of the same species, but Charlie insisted that the yellow one was best. Gene has succeeded in identifying the natural as Melanoplus Differentialis, a common type, after Thomas.

I did not use Charlie's imitation that evening, but I now have every reason for remembering that particular gift for a long, long time to come. It is no longer in my possession, being now the property of one of the legendary trout of the Letort unless it has rusted away and  fallen from that capacious mouth. I got into him several days later, when he was gorging on the natural, gulping and slashing at every one that came down his line of drift. The grim contest that ensued lasted for two hours and a half, necessarily so, for I was using 4x gut and had to let him have his way in order to conserve my tackle.

He was a tremendous fish, awesome in point of length, girth, and dignity. Crisis after crisis took place during this epic struggle—times when he dashed through one weed bed into another, my leader dragging heavily from the burden of weed caught by this maneuver, other times when he darted in one direction or another, looping the line around an obstruction by making a wide turn and returning to his original position. Then there were times when he revolved interminably in a tight little circle, slapping viciously at the leader at each turn, forcing me to gauge to a nicety the exact psychological moment when the leader should be slackened in order to soften the blow. All of these emergencies I was able to meet with proper countermeasures, but he finally broke me after the leader had been worn thin at the jaw line and when he was wallowing hoglike in a little bay where I expected to beach him. I had to stand helplessly, watching him feebly inch his way into the current and safety, his head lolling and his massive tail arching slowly from side to side. I dare not guess at his weight for fear of inviting the questioning glances of my polite friends. In any event I have a reliable witness to that memorable engagement, one who will swear to at least one-half of what I say!

I found little comfort in the proffered condolences of my sympathetic companion and I could only increase my discomfort by recalling a similar incident which befell me one clear evening, years ago. The memory of the awful thing that rose and engulfed my fly at the tail of a long pool and the subsequent break was considerably sharpened by my present loss of the Letort giant. My musings drifted into speculations about this and other trout which have become a substantial part of the legend and folklore surrounding this amazing little stream.

Could my fish have been that monster that rose,

wraithlike and sinister, from the murky deeps and hung suspended and motionless on the surface, his huge dorsal fin protruding from the water, before the pop-eyed and incredulous gaze of farmer Jones, who was walking his pasture at the time? And did he know, by his wisdom acquired from contentions with mankind, that farmer Jones was seriously debating the choice of pitchfork or rifle to end his existence, when he sank out of sight at the critical moment?

Could he have been that other monstrosity before whom a lady fisher, that Diana of the Letort, dared to cast her feathered offering, which was accepted, thereby making her an unwilling and tremulous appendage while he towed her slowly but relentlessly downstream for two solid hours, finally leaving her in tearful despondency?

Or was he that appalling creature that I saw from a high bank, one clear summer day, my footsteps sud-
denly halted and all motion stilled by the apparition of a giant trout slowly easing his way through the clear waters of the Letort? I remember thinking at the time that a yardstick would not measure him. Here memory cannot go much further, for there was only that one terrifying instant when he rose slowly and majestically, looked at my fly and refused it, then gently sank out of sight forever. Few of us have enjoyed the great good fortune that attended Don Martin's glorious achievement when finally he landed his stream-bred 15½ pounder at Newville.

With good fortune or bad, a single incident of this nature, necessarily rare, does not establish conclusively the excellence of the aforementioned grasshopper pattern, but it goes a long way toward convincing me that it has merit and should serve as a model whose virtues must be copied.

But many pitfalls await the man who is bent on dis-

41

covering and imitating all of the insects upon which trout feed. I venture to say that of all the insects which inhabit our streams none are more plentiful or more freely taken than cress bugs, Mancasellus Brachyurus. They exist and multiply in unbelievable quantities in these waters, harbored in comparative security by the dense weed beds of elodea, and are available to these trout any day, any season of the year. They are truly the bread and butter of trout diet in these limestone waters.

Consider the following seemingly implacable logic. If this insect is the most plentiful of all and if this is the one the trout take in greatest volume, *ergo* it should be the one to imitate above all others and, what's more, should occupy the most important position in a new series of artificial patterns.

Who would censure me for following this line of reasoning? Who would dispute the propriety of trying to imitate this interesting creature? What angler flytier would not rejoice at the opportunity of expending time and labor on such an imitation, with the prospect of enjoying marvelous sport with it even though it must be fished wet style? These were the considerations which prompted me to try to imitate the cress bug, but believe me, abstract logic of this kind, without practical considerations, sometimes commits us to the worst kind of folly. Rest assured that more time and effort were spent on this one insect alone than on all of the others combined, and always with the same negative results. Repeated failures with the various concoctions only served to dampen and finally destroy the earlier confidence with which I began this task, and everywhere I went, limestone anglers increased their demands for a successful pattern of the cress bugs. They, too, were afflicted with the same kind of logic that guided my investigations.

For a long time, longer than I care to remember, I

was not aware of the hopelessness of trying to imitate Mancasellus, I clung to the belief that it was only a matter of gaining the right size, color, shape, and translucency. But all variations in these respects were worthless.

The reason for failure did not lie in the mechanical execution of the imitation. The reason went much deeper —in fact, into the darkest recesses of the weed beds. For I finally discovered that Mancasellus is not a free swimmer. It is rarely found in mid-water, never on top, occasionally on the bottom. Hence the trout never get it except by burrowing into the dense weed beds and picking it off the stalks and branches of elodea, chara, and other weed growth.

It must now be clear to everyone that to cast an imitation of Mancasellus, no matter how perfect, into the weed bed itself in order to attract a feeding trout is worse than trying to find the proverbial needle in a haystack; for in the latter case there would be but one needle but in the former, countless numbers of the living likeness of the imitation that no trout could possibly find.

At this point an interesting question must have occurred to the reader, as it did to me. If a good imitation of Mancasellus were to be cast, not into the weed beds, but to a trout feeding in mid-water where his vision is not obstructed, would he not recognize and accept it confidently as the kind of food that he knows best? Unfortunately, the answer is no! Because of the physical characteristics of these waters and the feeding habits of the fish, the trout are rarely found in mid-water. The dense weed beds of elodea fill the streams from bank to bank and the growth almost reaches the surface of the water in midsummer, but there is always a very deep but narrow channel where the main current runs, and sometimes multiple diverging channels running parallel to the main one. The trout habitually lie in the bottom of of these channels, whether feeding or at rest, and rarely take a feeding position over a weed bed itself. To do so would be an assumption of risk that these wild creatures cannot afford, since the water over the weeds appears to be very shallow and provides no cover. If these fish are inclined to feed, they must do one of two things—take their food off the weed bed itself, or come to the top for surface food. Confinement to the deep, narrow channels does not allow for much wandering about for mid-water foods, and besides, this kind of effort is not necessary. But I must say, to the unending delight and satisfaction of all those who fish these waters, the trout often prefer to come to the top when surface food is present. This conclusion is inevitable since we know that Mancasellus is always present and easier and safer to secure. The reason for this preference is not difficult to determine when one considers that the cress bug, in spite of its large size, is for the most part composed of a hard, shell-like covering and very little meat. On the score of food value, he is really second choice to anything else on the surface and certainly not to be compared with the fat duns, grasshoppers, Japanese beetles, and the like.*

I earnestly hope that this account of the cress bug will serve to dissuade all of my brother anglers from any further attempts to employ or manufacture imitations of

* These conclusions are offered as perfect examples of the qualitative preference which trout seem to exercise when there is a choice of foods. I am fairly certain that the quality of various insect foods is the most important factor in explaining the selectivity of feeding trout. Such selectivity can exist between foods above and below the surface as well as on the surface. It does not really make any difference to a trout whether his food is wet, dry, or slightly damp; it is likely that he never knows. These distinctions exist only as a procedural index to which the fisherman must constantly refer in order to reach the level of the feeding fish with the proper lure.

A rare sight: a big brown trout poised for a rise in shallow water over a weedbed. Big trout are sometimes discovered this way early in the morning, after an all-night feeding.

The savage, slashing strike of a wild six-pound native Letort brown trout. The author worked for hours in a blind to capture this fine shot at just the right split second.

this insect, for such imitations are bound to be worthless in actual use. This I must emphasize in spite of the occasional success that some anglers enjoy with imitations that they are pleased to call cress bugs. In any event these successes occur when the imitation is used on certain gravelly stretches, free of weeds, where the artificial of Mancasellus ranks no better than any one of the modern nymph artificials.

There are other peculiarities in the character of the water and the feeding habits of these trout which must be considered in devising new patterns. The placid surface, the gentle current, and the clarity of the water form a combination that bodes ill for the angler whose fly pattern is not correct, for the trout attune themselves to these conditions and reflect all of them in the deliberateness of their movements, the careful inspection of each passing object, and their surreptitious manner of actually taking food from the surface. Sometimes the take is so nearly imperceptible as to seem doubtful to the keenest of eyes, and the education of the limestone fisherman is never complete until he has come to see, recognize, and catalogue in his mind every little disturbance of the surface that spells trout taking top-water food. He must learn  to know that tiny flick, that little curl of the water that appears and is gone in a flash. Sometimes it is a little roll, or humping of the surface, but he must never, never mistake this for a nymphing trout, as we noted earlier in this chapter. This is a true surface rise and the water is, in fact, broken, although the angler will probably never see the actual rupture of the surface.

It is a peculiar sensation to anyone who has a sound knowledge of these things to sit and watch a phenomenon of this nature while a parade of anglers go by, never seeing, never knowing that over against the far bank a lusty three-pounder is feeding industriously at regular intervals for perhaps four or five hours.

Even though armed with this knowledge, the expert is oftentimes unaware of the presence of surface-feeding trout. The following incident is appropriate for the purpose of illustration: three veteran limestone anglers sat on the bank of the Letort on a day in mid-July during the 1946 season, not more than twenty feet from the nearest edge of the stream. They had been there for several hours, chatting, smoking, and watching the water carefully. One of them finally noticed a small disturbance of the surface against the far bank. A short interval elapsed and again the disturbance appeared. The angler retrieved his tackle, made the proper approach and sent his fly to the desired spot. A fine fish of some 2½ pounds was raised, hooked, and landed and an immediate autopsy was performed. All of the stomach contents were perfectly recognizable. Digestion had not taken place to any degree and the following food forms were revealed: 15 or 16 ants, 5 or 6 houseflies, 3 or 4 lightning bugs, 2 cress bugs. The alternative figures represent allowances for broken parts. It is significant that this trout had been feeding continuously in the presence of these very keen dry-fly anglers, and it is more significant that all but two of the food-forms were of the surface variety; and yet this activity went on unnoticed for a considerable length of time.

Is it any wonder that such trout are not to be taken so easily! More than this, however, is required of the angler who seeks the maximum in sport with trout of the limestone waters.

The matter of proper presentation of the fly is a thing which really belongs to a book on the art of fly-casting. That subject is not within the province of this volume, but surely some reference to this subject must be made herein in order to insure good results from the new patterns; for no matter how perfect an imitation may be, if it is not delivered in the proper manner it is bound to fail, along with the reputation of its designer.

The matter of presentation calls to mind another peculiarity of Letort trout, in particular, with reference to the manner of taking the dry fly; and must be divided into two distinct phases: (1) the observation post, and (2) the taking position. In the vast majority of cases, these trout do not lie close to the surface at feeding time but occupy a position very deep in the channels, as noted before. When the fly is pitched—note carefully—in front of the observation post—not the taking position—the first movement is a gentle swaying motion rearward, utterly graceful and continuing low in the channel for a distance of some two to four feet; then there is a sharp lift of the body upward and the fly is intercepted at the taking position.

Time after time I have commiserated in silence with a brother angler who persisted in casting to a taking position, not realizing and not appreciating the fact that he was casting his fly, very accurately, let it be said, but uselessly to a spot some three or four feet behind the trout; and they simply will not come back that far to take a fly. This is an excusable error, for it is given only to the most rigidly trained and keenest of eyes to know that the shadowy detachment occurring in the depths of of the channel is a trout moving backward under the fly.

On the other hand, it sometimes happens that the trout will often halt the sharp upward movement just short of the fly, reverse himself and follow the fly for three or four more feet downstream, no doubt for a closer inspection. Nothing is more trying to the angler than this exasperating habit; for, although the threat of drag in a normal drift of three or four feet is worry enough, when

47

it is stretched to seven or eight feet the problem is considerable. There is always impending the crisis when free float ends and drag begins, and the trout may have chosen that moment to decide that the fly is authentic. The result in most cases is a disdainful flick of the tail as the trout drops low and slithers back to his original position, the observation post. If the unfortunate angler encounters this misfortune often enough, it begins to assume a nightmarish quality, to such an extent that it will disturb the composure of the sturdiest of natures, not infrequently manifesting itself in actual physical tremors.

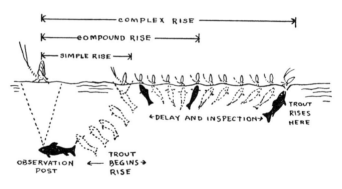

There is yet another variation of this feeding habit which the angler must recognize in order to present the fly properly. Sometimes the line of drift flows straight downstream, then angles sharply away toward the center of the stream; this is often caused by an obstruction near the bank, a weed bed, or a bend in the stream. Sometimes the change of drift originates in the center and is directed toward the banks. In either event a trout may occupy an observation post at the exact point where the flow begins to break away. In these cases, the first movement of a trout is not directly backward, but consists of a gentle, undulating movement sideway, pro-

48

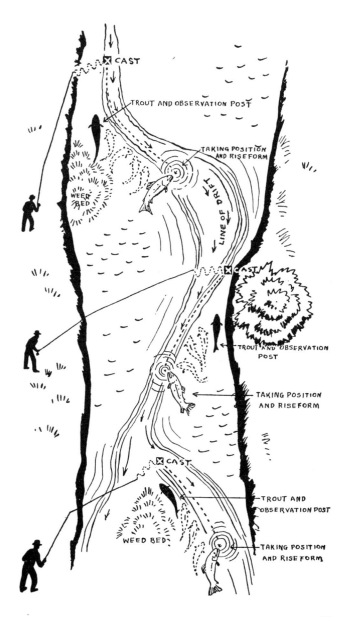

CAST

TROUT AND OBSERVATION POST

TAKING POSITION AND RISE FORM

WEED BED

LINE OF DRIFT

CAST

TROUT AND OBSERVATION POST

TAKING POSITION AND RISEFORM

CAST

TROUT AND OBSERVATION POST

WEED BED

TAKING POSITION AND RISE FORM

49

gressing across the stream and low in the channel for three or four feet; then comes the sudden, sharp, upward lift and the fly is taken.

The unsuspecting angler may easily be deceived into casting his fly directly upstream, thinking that the observation post is ahead of the taking position, when in truth the cast should have been made four feet to the side of the taking position and two feet ahead.

The case of the box-elder trout is a graphic illustration of this point. I first saw this fish one sunny afternoon in mid-June. His taking position was four feet below the trunk of a box elder and four feet out from the bank. I watched him feed for a while, then decided to try for him. When a number of casts proved of no avail, I ceased all further attempts until I could make a study of this particular fish. On my next visit to this spot, I was elated to discover that he was feeding in his usual quiet, unobtrusive manner. A study of the current at this point revealed that the line of drift was concentrated at the trunk of the box elder, then broke sharply away toward the center. I felt sure that this trout's observation post was some four or five feet upstream and very near the trunk of the box elder, thereby requiring a cast a foot or two above the trunk and a float of six or seven feet in order to take him. I could not have picked a more unfavorable day for this trial. A westerly wind was steadily blowing half a gale directly cross-stream toward me so that in order to reach the desired spot with my fly it was necessary to shoot the cast through an opening in the branches of the box elder some three feet in diameter. Time after time my casts were buffeted about, blown back and misdirected. My fly landed on every spot near this trout but the right one. I stopped casting entirely, gently waving the rod back and forth, hoping for a cat's-paw in the wind in order to make at least one accurate

50

The Letort—at the famous barnyard stretch where many noble battles were won and lost.

Cedar Run—a little gem of a limestone stream.

cast. A momentary lull finally came along, and seizing the opportunity I shot the line through the opening with all the speed I could muster. I was rewarded with a perfect cast and almost immediately I could barely perceive the shadowy outline of the fish as he moved sideway and slightly backward to intercept the fly. The surprise engendered by the unexpected success of the cast and the actual rise of the fish induced a momentary loss of poise, causing my hands to freeze to the rod and line while the trout made a sudden lunge and broke me. This same fish, a fat three-pounder, was subsequently captured by Charlie Fox in his usual skillful manner and exhibited evidence of a fresh scar on the lower left jaw.

For the above reasons, I will continue to advocate the use of slack-line casts rather than the curve casts, wherever the former are possible. The great fault with the slack-line casts lies in the fact that there is the likelihood of lining the trout with the leader, directly overhead; but this can easily be avoided by making the casts at right angles to the stream or nearly so. The long drift afforded by a slack leader will oftentimes overcome the difficulties of the delayed rise.

The application of the slack-line delivery to a variety of circumstances makes it especially valuable. Let me mention another of these situations which is common on limestone streams.

Let us suppose, for example, that the angler is standing on the left bank, looking upstream. Directly across-stream, against the far bank, there is a small bit of backwater or bay in which a trout is feeding very delicately on tiny duns, but feeding so close to the weeds which border the stream that part of the rise-form itself is dissipated against the weed stalks. The feeding trout seems to be actually touching the weeds in these instances as he takes in one insect after another, and usually he will not move a single inch from this line of drift. Let us suppose further that there is a narrow piece of fast current between the angler's position and the little bay or backwater. Obviously, the caster's problem is to place the fly tight against the bank of weeds without actually touching them but not more than a half-inch or one inch away from them, at the same time insuring a long free float to circumvent the possibility of drag by the intervening current. Curve casts are absolutely useless in these cases since the curve itself lies in the slow backwater, affording no pressure against the upstream side of the curve to equalize the pace of leader and fly. What really happens is that the line is immediately seized by the current, causing leader and fly to describe an arc, actually traveling upstream, then they are whipped around and taken by the fast water out of sight.

To my knowledge, there is no situation more exasperating than this one, especially if the angler allows his determination to exceed his patience. Time after time, ignoble defeat is the lot of expert and novice alike under these conditions and the temptation to abandon these fish entirely and forever is common to both. However, the angler need not despair of ever taking these fish, for there is at least one maneuver that is successful part of the time. It is questionable that anyone will ever score heavily on fish of these habits, but if the angler wishes to try he can employ a special slack-line cast which is executed according to these directions.

Let the angler take his position directly across from the trout in such a way as to face the opposite bank squarely. Aim the cast high above the weeds and deliver it far beyond the trout, *over the land* on the opposite bank for perhaps ten feet, bringing it to a halt with a decided abruptness that will cause it to recoil sharply toward the angler, thereby achieving a series of close-packed loops on the surface of the backwater and allowing the fly just barely to clear the weeds on the return and fall within an inch or so of the weedline. Theoretically it is a successful cast, since the series of loose coils pay out slowly into the fast current, allowing the fly to pursue a natural and leisurely drift in the backwater. In actual practice, the placement of the fly within an inch or so of the weedline by this indirect method is a rare bit of luck even the most expert would appreciate. Of course, this procedure is unnecessary if it is possible to cross to the other bank where a short line can be cast with ease to such a fish, but some of these streams, particularly the Letort, are too deep to permit such a crossing.

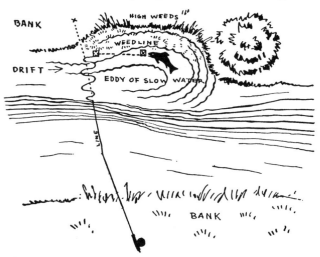

One more observation in connection with this cast is worthy of note; that is, in order to obtain maximum benefit from the slack line, the forward cast must be delivered with considerable force, stopping its forward motion sharply, thereby causing it to jerk backward and fall in loose curves on the water. I have never been able to do it properly with soft rods and light lines. In recent times, I have resorted to heavier lines and stiffer rods in order to accomplish my purpose. During the season of 1947, I acquired and used exclusively a fly line with a belly dimension of .070 inch, or triple A. This was variously described by my friends as clothesline, bull rope, lariat, and so forth; but I can truthfully say that I never enjoyed a more successful season with these difficult fish, and that I ascribe a great measure of this success to the use of this combination, simply because I can throw a better slack line.

Such is the character of the fish and the fishing on the Letort and other limestone waters—exacting to a high degree, where tackle and technique must be finely adjusted and, above all, the artificial pattern must be correct. This last requirement is always uppermost in the minds of those who practice the dry fly on these waters; and any effort to improve patterns has always met with the approbation and encouragement of these fine anglers.

It is the natural consequence of adherence to the principle of *following the hatches,* systematically pursued from date to date in accordance with the tabulated emergence periods of the important insects. It means more than fishing to all of the hatches of a single stream, good and poor, for the entire season. It means, rather, a constant pilgrimage from stream to stream at the appointed time when the heaviest hatch of fly for each of them is in progress or about to begin. I do not know how much this principle is followed elsewhere, if at all, but for these anglers of the limestone streams it is more than a habit; it is a religion.

Little wonder, then, that for them the artificial has become an object of special concern, suffering as it does the eternal competition of the naturals for the attention of the discerning trout.

54

This competition has caused the dry-fly angler of these waters to look hard and long, now at the artificial, now at the natural, bringing into sharp focus sometimes the merit, sometimes the deficiency of the dainty confections that we call dry flies.

Chapter 3

FLY-DRESSING THEORY, or the Shape of Things to Come

TALK about the dry fly must always begin with Frederic M. Halford, first historian and high priest of a puritanical cult which has spread its influence over the entire world of the fly-fishing fraternity. Gifted by nature with an analytical and inquiring mind, he was well fitted for the task of assembling in orderly and lucid manner all of the theory and practice of the dry fly as it existed in his day. His gifts extended to embrace not only the investigation and solution of angling problems but also a mode of expression which was well suited to the task. Whether he was right or wrong in his conclusions, there is never any doubt about his meaning, and for those who delight in the scholarly view there is a peculiar satisfac-

57

tion in following his well-balanced, well-constructed sentences and paragraphs. His diction is faultless and adroit and best of all, he studiously avoids the use of colloquialisms which might have lost their meaning to modern readers.

There is somewhat of a mystery surrounding the writing of Halford's first books. It seems, according to many of his remarks, that he had originally planned to write them in collaboration with his friend and mentor George S. Marryat, whom he regarded as the finest fly-fisherman of those times. This opinion of Marryat was shared by everyone who wrote about him, and somewhere in those writings can nearly always be found an allusion to him as the Master. His skill and accomplishments as an angler became legendary and his character was no less admirable.

By Halford's own admission, he became heir apparent to Marryat's great store of angling knowledge; in his earlier books, he intimates perhaps that in many parts Marryat is speaking to the reader through Halford. Nowhere does this impression become more pronounced than in Halford's autobiography.

Why Marryat himself did not write, if not the first, at least one book on the dry fly has been a matter of speculation among many people, and why he declined to collaborate with Halford according to the original plan only serves to increase this speculation.

It is the general consensus of opinion that Marryat could have written a great book, but nowhere does there seem to be any confirmation that he had the ability to do so. In all likelihood he understood better than anyone else that a remarkable historical event was about to take place. The treatise presenting the revolutionary dry fly to an unsuspecting public was not to be undertaken lightly and certainly was not to be ushered into being by ordinary talents. The inference is strong that Marryat appreciated Halford's consummate skill of expression, which undoubtedly he valued far above his own ability or that of his other contemporaries. It is logical to believe that he, reputedly generous and self-effacing by nature, would deny himself or anyone else the public acclaim which would fall to the man who made the first pronouncement, unless he were fitted to be the bearer of glad tidings.

That Halford was fitted for this mission there is now no doubt, and if this was Marryat's judgment he was amply vindicated by the enthusiastic response of the public to the first two books, best expressed, I think by G. E. M. Skues, brilliant writer himself, who described them as "hypnotic, submerging."

It would be entirely unfair to say that Halford depended completely on the knowledge and work contributed by others. Alfred Ronalds was his inspiration, Marryat his teacher, but the latter's early death and the limitations of the former left him without restraint in a field where originality of ideas was not yet by any means exhausted.

Thus in the last years of his life he offered to anglers his *Modern Development of the Dry Fly*, 1913, no doubt intending this work to be a climax of perfection to a long and pleasant career of his loved avocation, dry-fly fishing.

His earlier works, *Floating Flies and How to Dress Them* and *Dry-Fly Entomology*, contained and recommended respectively 90 and 100 patterns, representing, in all probability, nearly all of the dry patterns conceived and used by chalk and limestone stream fishermen of that period.

In his later years Halford realized the absurdity of the multitude of then existing patterns and sought to reduce

59

them to a reasonable number of useful imitations. *Modern Development of the Dry Fly* was the culmination of this effort and served as a vehicle to announce to the world of fly-fishermen his theory of strict imitation. This was the immediate objective but, more than this, it expressed an entirely new theme—the organization of the dry fly.

It was a systematic relationship of the artificial to the natural, and the significance of this fact was not lessened by the questionable value of Halford's method of imitation. It was a momentous and far-reaching step, and its influence will assert itself in every system of fly dressing, no matter what variations there may be.

Whether or not he attained a perfect imitation of the natural became the subject of a great controversy occurring chiefly in the works of the English writers who followed him, for the dry fly was virtually unknown in America and the rest of Europe at that time. Theodore Gordon, who godfathered the dry fly in America, made no significant changes in its composition, although his interpretation of wood-duck wing was somewhat novel and remains a favorite today. George LaBranche was using the dry fly but applied his talents mainly to the manner of its employment on our streams. Emlyn Gill simply described the English system and advanced his "checkerboard" method of fishing our waters. Louis Rhead's formal publication, *American Trout Stream Insects*, 1916, was an attempt to perform the same service for American anglers as Halford did for the English, a cataloguing of American insects and the creation of their corresponding imitations. It was the first American product along these lines.

Anyone who has read and studied Rhead's work cannot help but experience a feeling of regret and frustration, for it could have been the finest thing of its kind and an

60

invaluable aid to the angler-entomologist. The hand-colored plates of insects are excellent, for Rhead was a fine artist. Unfortunately he used a system of nomenclature that was confusing and unreliable and he included many insects that are of no particular interest to the angler—insects which never appear in great numbers or which are not taken by the trout, at least in the winged form.

Finally, Rhead positively discounts any value that the imitation may have as a dry fly and argued that "for every insect a trout takes alive at the surface, a thousand are consumed drowned underwater or near the surface; and to one natural insect able to float on the surface, there are hundreds which cannot float." That was a monumental error—an error which was compounded by Rhead when he used the floating duns as models for dressing his wet flies.

Nevertheless, it must be said for Rhead that he was a very keen observer and took note of many important things about floating duns, among them the fact that these insects hold their wings upright and close together, so that they appear as a single wing. Rhead promptly incorporated this feature in his imitations, always employing a pair of matched body feathers with concave sides together. The result is an abomination to cast, causing the fly to spin and the leader to twist. The effort is always accompanied by a fearsome noise, which is, to say the least, very annoying to the angler. The standard split wing does not offend in this way.

RHEAD'S FLY

Rhead also insisted on using a long extended body, i.e., detached, which undoubtedly aggravates the above-mentioned condition besides unduly weighting the fly, thereby making it almost impossible to obtain a proper float. All of these are details of construction which Halford eschewed at a very early date, discovering from his

61

familiarity with the dry fly, no doubt, that these features were undesirable.

In other respects Rhead's methods bear a striking likeness to the methods of Halford, particularly in the choice and treatment of materials, and it is a likeness which persists to this day wherever a specific imitation is attempted. The influence of Halford is still strong and it persists, as well, among many of the neo-imitationists whose change of viewpoint might have evolved something different, at least in the form of the artificial.

As a rule, the tying procedure follows the same general routine, with little variation. According to Halford's creed, the smallest details of the fly's construction were all important. Every part (except one) and the exact shade of every part of the natural must be represented. Every detail of the natural fly's anatomy must be included, no matter how absurd the result.

In carrying out his plan he went to great pains in fulfilling these requirements. Colors were selected and matched side by side with the natural insect with the aid of a powerful glass, the eyes were imitated by means of several turns of the proper shade of horsehair at the head of the fly, the segments of the natural fly were counted and the same number of turns of ribbing were included in the artificial.

It is almost certain that few amateurs or professionals ever attempted such meticulous exactness in color. For the amateur especially there is an almost insurmountable obstacle in obtaining the correct colors for all of the materials involved, which were required to conform to the specific shades of a color chart. Even for the professional dyer it is no easy task to obtain some thirty-five or forty shades to match the color blocks exactly. Here, however, any dissimilarity ends. Wings are tied in, now as then, immediately behind the eye of the hook, and are usually constructed of starling, duck, wood duck, or hackle only. Bodies occupy the remainder of the shank and tails are added in continuation of the body, serving mainly as an additional support on the surface of the water. Hackles are used to imitate legs and provide the main support directly underneath the wing.

CONVENTIONAL

Historically speaking, the dry fly is an adaptation of the ancient wet fly of Dame Juliana Berners, Charles Cotton, and more recently Alfred Ronalds. It never really enjoyed an independent orgin; the significant differences in the dry are not so much an attempt to imitate the floating dun as they are to make the wet fly a floater. As a matter of fact, the earliest dry-fly fishing was practiced with wet flies which were "cracked to remove the super-abundant moisture." If the dry fly had ante-dated the wet, it is highly probable that the structural characteristics of the two forms would be entirely un-related and that the dry fly would be based solely on observations of naturals. The development of the wet fly itself is the most illogical thing in angling history, since the wet fly does not represent an attempt to copy under-water insect life. The modern nymph artificial is a far more reasonable imitation and deserves to be classed as the only true wet fly.

It is plain, then, that a close scrutiny of the natural dun was never any real basis for dry-fly construction. A systematic and pointed inspection of the live insect on the surface of the water will reveal a number of irrele-vancies in the standard pattern which seem to have been ignored or but partially observed, leading one to believe that a more rational design could have been engineered in accordance with the performance and appearance of all classes of insects on the surface of the water.

Of course, dry-fly construction is necessarily a matter of many compromises brought about by the limitations placed upon the flytier by a number of unavoidable obstacles, and the inevitable hook is one of them. Practical considerations involving the durability of materials and their availability are others, and tying procedure must be kept within the bounds of ordinary skill.

Nevertheless, they do not constitute a bar to the improvement of fly patterns, for it is possible to accomplish the necessary changes with ordinary, easily available fly-tying materials without having to resort to complicated or highly technical preparations. Nothing tends to greater discouragement for amateur and professional alike than to be faced with a complicated tying technique requiring specialized or unobtainable materials; once attempted, such a system is quickly abandoned, and allowed to remain as a mere abstraction in the history of fly-fishing. It is an error which is common with the Halford and Rhead techniques and, more recently, the J. W. Dunne series of flies, and it is a fault which should be studiously avoided now and for all time in whatever theory is advanced.

The fly-fisherman and flytier should remember above all that the artificial is nothing more than what is intended, an imitation. It can never be anything more than that, and if the trout can perceive a viable distinction between artificial and natural, as he probably does, it is a gift of law to which he is entitled. These reasons suggest that the proper approach to discover the requirements of the artificial should stem from considerations of practicality in dressing and in use, the appearance on the water from the trout's point of view, and a careful review of the living insect itself, terrestrial and water-borne.

"What the fish sees" is a topic which has acquired a nature so delicate, so charged with controversy that the mere mention of it will evoke violent thoughts and expressions. Some are disposed to dismiss it summarily, saying that it is impossible to see with the eyes of a trout. Others are inclined to the belief that the trout sees things very much in the same manner as humans. These are extreme views and can hardly be correct particularly in view of recent scientific findings revealing much of the nature of a fish's eyesight and certain peculiar physical laws which affect the appearance to the trout, of objects in and out of the water. Halford and his contemporaries never mentioned or applied such laws although earlier, Ronalds wrote sparingly about the principle of refracted light but only as it affected the fish's view of the angler. Modern anglers cannot afford to disregard these new findings especially because they suggest important changes in dry-fly dressing.

How different is the tone of modern writers who venture on this subject, lacking entirely the resolute and uninhibited style of Halford and the early authors! One and all they exhibit a characteristic hesitancy and indecision, born of a painful awareness that much of this matter lies in the realm of conjecture and is subject to revision owing to the embarrassing perversity of trout conduct.

Yet in that way lies real progress, for here and there a solid fact is exposed, now and then a worthy conclusion is attained, that can be demonstrated with mathematical certainty. These we must accept in so far as they form the prescription for a logical dry fly; when more is added to the store of knowledge, the prescription must be altered to conform accordingly.

The task of discovery belongs primarily to the province of the physicist, the logician who assembles and interprets the observations, and the artist fly dresser who

65

executes the logician's directions. It is possible that one man alone may possess these qualifications.

Beginning in 1910 with *Marvels of Fish Life* by Dr. Francis Ward, followed later by *Animal Life Under Water*, 1919, the first complete treatment of the trout's point of view was brought to the attention of the fly-fisher. In 1914 J. C. Mottram's *Fly-Fishing: Some New Arts and Mysteries*, suggested some interesting applications of new principles in the art of fly dressing. In 1924 J. W. Dunne's *Sunshine and the Dry Fly* offered an analysis of color from the trout's point of view and a novel way of obtaining translucency in dry flies. The most comprehensive of all is Col. E. W. Harding's work, *The Fly-Fisher and the Trout's Point of View*, 1931, wherein the broad outlines of the whole subject are drawn for the purpose of study now and in the future.

Although these are the works of English writers, serious attention to the fly-fisher's problems has also been given by a few Americans, notably Mr. Edward Hewitt, whose several books are worthy of anyone's reading. All of these do not necessarily tell the whole story, but they represent, at least, a solid foundation for the neo-imitationist who must make a beginning with the accumulated knowledge of others before he can make a contribution of his own. That is the story of the fly dresser's progress.

Subjectively, it is a difficult matter to say how a trout sees things, since nature has probably created physiological changes to adapt his eye to view things through a medium other than air, namely, water. Objectively there are a number of optical facts, easily established, which govern the manner in which a trout sees things; one of the most important of these is the existence of the trout's "window," that porthole in the water through which he sees the world above and objects on the surface. The formation of the window is the result of well-known physical laws concerning the manner in which light rays are bent or refracted when entering the water.

Elsewhere in angling literature, to which allusion has already been made, can be found an exact scientific exposition of this phenomenon. The restatement which follows is submitted for the purpose of relating the important facts to the requirements of the dry fly. Even though the analogies which are drawn may seem a little weak in some respects, they may serve to provide a more understandable explanation to many of us who are not familiar with the terminology of the physicist.

Light rays penetrate the water from all directions except where the rays strike the surface at an angle of less than 10 degrees to the surface. It means that rays which travel almost parallel to the surface are reflected upward and hence do not reach the eye of a trout. It is somewhat similar to the result of throwing flat stones along the top of the water in order to make them skip. Every boy who does this eventually learns that the more nearly parallel to the surface the stone is thrown, the better it will skip. When the stone is thrown at a steep angle, it simply enters the water without rebounding skyward. Once having entered, however, it does not continue in a straight line to the bottom from the starting point; it encounters more and more resistance, losing velocity on that account, and falls to the bottom at a steeper angle than when it entered the water. If the stone is not thrown at an angle but allowed to drop straight downward to the water, in other words a perpendicular or free fall, the path of the stone to the bottom is likely to be a straight line from the starting point. Rays of light act in a strikingly similar manner; like the stone, when they begin to strike the water at an angle which is steep enough, they start to penetrate and, having penetrated, are deflected downward at an angle steeper than that at which they entered.

The minimum angle at which they begin penetration has been determined as being about 10 degrees, and the angle at which they slant toward the bottom has been measured and established at 48½ degrees. Entering as they do from all directions, the rays which begin to penetrate at the lowest angle of 10 degrees form the outside limits or perimeter of a circular area through which a trout sees the world above and beyond the water. Furthermore, these same rays which mark the limits of the "window" and slant downward at an angle of 48½ degrees from all sides must come together at some point below the surface and form an inverted cone or funnel-shaped area of visibility. The position of the trout's eye corresponds to the small end of the funnel.

Wherever the trout moves his window moves with him, because his eyes continue to intercept a similar group of rays no matter where he may be. It is like being wet with those raindrops only which fall where you are standing; move to a different place and you are struck with a different series of drops, but no less wet.

It should be clear, then, that a trout is committed by natural laws to a definite angle of vision; however, he may enlarge or diminish the area of the window by moving up or down as he pleases. If upward, the window becomes smaller, since the diameter of the base of the inverted cone is shortened, the angles remaining the same. If downward, the area of the window becomes larger in accordance with the widening of the base, again the angles remaining the same.

Beyond the circumference of the window the surface of the water is hidden, and the underside of the surface appears to the trout as a mirror which reflects the bottom of the stream and objects in the water. Although the surface is invisible, it is possible for him to see objects on the surface, beyond the window, which are high enough to intercept those rays of light which enter at the edge of the window at an angle of 10 degrees or more. We have generally lost sight of those lesser rays which are totally reflected from the surface, and that is a mistake. They perform a valuable function, too, for they serve to illuminate by reflection all objects coming into the window. In this way a trout gets a sort of cross-illumination somewhat analogous to the photographer's arrangement of lamps. It is also a compensation for some of the loss of light because those rays which enter near the edge of the window are partly reflected, but that loss is of small moment since the light rays at that point are more numerous and crowded together than anywhere else in the window. It is important to know, however, that the mirror, sometimes called the area of total reflection, has significant meanings for the dry fly as well as the wet fly.

It is an elastic thing which bends and receives impressions from the weight of objects on the surface beyond the circumference of the window. These impressions are conveyed directly to the trout through the water even though he cannot see the object which makes them. The  feet of a floating dun, pressing upon the upper side of the mirror, create small meniscuses or lenses which rise about the feet and encircle them to form light condensers arranged in a definite pattern. This is the "light pattern" which is variously described as the first indication to the trout of an approaching fly. It is important to note, too, that the formation of this pattern varies considerably with different classes of insects. If a trout is induced to begin his rise by the recognition of this repeated and familiar index, it is absolutely necessary that

69

the prescription for the artificial should begin with an attempt to imitate the appearance of the natural *before it enters the area of the trout's window.*

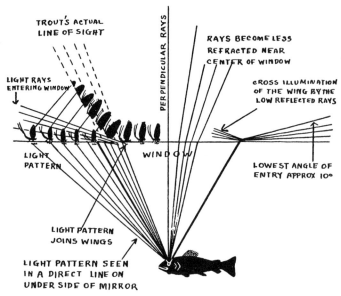

Normally the drifting dun stands upon the water, touching it only with the points of its six feet arranged in parallel lines of three. The body is curved upward, free of any contact with the water, with stylets pointing to the sky and never on the water unless the insect is injured or waterlogged. In the case of some insects, especially the heavier mayflies, the thorax may touch the water and the light pattern is changed to that extent. In extreme cases, the body is curved upward so sharply that it meets and blends with the rear edge of the upright wings and the insect appears as a composition of wing and thorax only.

Now it is hardly conceivable that a trout would quibble over a few points of light, more or less, in the light pattern, but there is a vast difference between the fly

70

whose body and tails have broken through the opacity of the mirror and that which indicates its presence by a few points of light only. It must be remembered that the floating dun, except for the indentations on the surface, is invisible when it is outside the window. On the other hand, the spinner or imago of that same dun lies with its body and tails in the surface film, partly submerged, and is visible to the trout when it is beyond the edge of the window. Strictly speaking, there is no such thing as a light pattern for the spinner, for the trout sees the insect itself through the water in a direct line. It is plain then, that the standard tie for the dun, with its tails and body resting on the water as they often do, is not an accurate representation at this stage. There is really no point in distinguishing between dun and spinner unless they are tied to conform to the separate habits of each of them on water, always assuming, of course, that it is a valid distinction for the trout.

SPINNER

As far as a trout is concerned, there can be no difference between a dun which is tied to drag body and tails in the water and a fly which is tied to represent the prone position of the spinner. Tank observations of the artificial fly have been offered as proof that the trout sees the body of an insect clearly but only because the artificial floated awash, like a sodden spinner, instead of a dry and light-riding dun. Let the living dun itself be viewed in the normal pose that it affects on the water. The difference from a view of the spinner is acute.

ANT

This distinction does not end with dun and spinner. There are other insects, no less important, whose aspect on the water demands a similar treatment. The large ants, black or red, for example, are partly sunken with wings awash. So, too, are the grasshoppers and the Japanese beetles. Yet the light-pattern theory is subject to some qualifications, tending to minimize its importance

GRASSHOPPER

JAP BEETLE

71

to some extent, in view of the fact that some insects upon which trout feed make no light pattern whatsoever. That is something which seems to have escaped the notice of so many observers and yet it happens very frequently, especially in the case of the minute insects. The very small ants, the tiny duns, and the smallest jassids, lacking any appreciable weight, make no indentations on the surface film that are observable; the actions of the trout positively confirm the truth of this statement. These are the times when the rise is very often delayed and the backward drift is prolonged, two indications that the trout has not seen the insect until it is in the window. What is more indicative, the utmost accuracy in casting is required to place the fly within the limited circumference of the window; else it escapes the attentions of the trout, particularly when the trout are hovering close to the surface, as they sometimes do when feeding on small insects.

Nevertheless, there are times when the light pattern appears to be the only and final inducement to rise; this is probably true, for example, where caddis flies are fluttering on the water beyond the edge of the window. The series of light flashes caused by the gyrations of this insect is undoubtedly an exciting factor. Thus the "bump"  cast of George LaBranche is a pure application of this principle, and the pattern of fly never really matters for this fishing so long as it can be manipulated to imitate the moving caddis. It is hardly possible that the caddis itself can be seen by the trout who must pursue this lively creature while diminishing the area of the window to almost nothing as he nears the surface.

Admittedly the light-pattern theory is valid and must be carefully observed as the initial step in the construction and delivery of artificial flies; yet there is a disposition on the part of some to overemphasize this element and to minimize the importance of the insect's appearance in the area of the window.

Moved by a powerful curiosity to discover more about the relative importance of the window and mirror, Mr. James Kell of our fly-fisher's company devised and finally engineered a plan during the height of the Hendrickson hatch on the Yellow Breeches a few years ago.

An opportune time and place were discovered where a fair-sized trout was feeding with clocklike regularity on the drifting duns. The trout's position was located in a narrow line of drift where he was able to intercept any or all that he pleased. After noting that the feeding trout allowed very few of the Hendricksons to pass, Mr. Kell proceeded upstream and stationed himself very close to the same line of drift that carried these insects to the waiting trout below. As the drifting duns approached, Mr. Kell began to pick them off the water and, holding them as delicately as possible, severed the wings from the bodies; he then carefully replaced them upon the identical line of drift and sent them to the waiting trout downstream, all of them still alive and able to assume the characteristic pose with body and stylets pointing skyward. Time after time this operation was repeated. Sometimes several wingless bodies were placed to encircle a winged natural and sometimes these were placed in a long breach between the winged naturals. In all, 37 wingless bodies were sent down over the same feeding trout *and not a single one of them was taken!* The winged insect, even when surrounded by the wingless bodies, was always chosen to the entire exclusion of the latter.

Now that is an earth-shaking experiment for the fly-fisherman and it contains many implications which cannot be idly dismissed but must be properly examined in order to form an accurate estimate of their worth.

In so far as the Hendrickson is involved, it is conclusive enough that trout do not accept it on the sole basis of a light pattern beyond the edge of the window. But this does not mean that the light pattern is unimportant for this particular species. Subsequent observations in clear water showed quite plainly that the light pattern beyond the circumference of the window caused the trout to begin his rise only to check his movement when the wingless insect entered the estimated area of the window.

In all of these instances the performance of the trout was consistently similar. In the beginning, when he is prompted to begin his rise by the disturbance of the surface film, there is an accelerated movement of the fins without more; then comes a gentle swaying movement backward for a short distance and finally a sharp upward lift which is suddenly halted a short distance perhaps a foot or so, from the surface. From these facts it is most reasonable to conclude that for some insects a light pattern is no more than an invitation to rise, final acceptance being conditioned upon confirmation which takes place by observation in the window.

What that confirmation is involves those further implications suggested above, and sheer logic urges that it is no more and no less than the presence or absence of wings, for they were the only missing parts.

This may come as a shocking surprise to a vast army of fly-fishermen who are inclined to deprecate the value of wings and who sometimes claim that the trout do not see them. It does not agree, for example, with the conclusions of Col. Harding, who maintained that wings are not as important as a light pattern except, perhaps, in the instance of the larger Ephemerids such as the Green Drake. He argued that the usual illustrations of the gradual appearance of the wings are misleading, since they

74

do not take into account the refracted height of the wings as they enter the window.

Assuming that he is correct, his calculations show that the large wings of the Green Drake are refracted to a height of only ⅓ inch at one inch beyond the edge of the window, hence quite unnoticeable. Yet a trout feeding on Hendricksons, a much smaller insect, would not be convinced!

In actual practice there is ample assurance that wings for floating duns are of paramount importance. There is, for example, the effectiveness of hackle patterns, which are tied without bodies or tails and the astonishing success of flies such as the Hewitt Spider, used by many limestone anglers during the hatch of the Green Drake at Spring Creek and other central Pennsylvania limestone streams. Hackle which is wound in a flat plane, as in the Hewitt Spider, forms a surprisingly good wing. Mr. Hewitt invented a better fly than he really knew! Here, however, we must enter a caution, for the Green Drake hatch is a poor standard by which to judge fly patterns. At such times trout are often caught by many strange enormities; so mad are they, once the feeding has begun, that they are incapable of the usual caution.

NEVERSINK SKATER

My earliest success with flies of this type was a source of wonderment to me for a long time. Once, in a fit of daring, I tied a few patterns of unconventional design to imitate the little pale watery of the Letort. This stream has always been a difficult one for me to solve and for many others too. The capture of one good trout in an evening's fly-fishing was quite an achievement. Two good trout was a memorable event. Three of them would establish an enviable reputation!

The fly in question is before me as I write. I have preserved it for many years to remind me of my ancient amazement. It has pale gray hackle with a tinge of honey, tied buzz but thinly, and no larger in diameter than the height of a pale-watery wing. The body is very short and stubby, constructed of yellow fox fur on a short shank No. 16 hook, without tail support.

On the evening of its baptism I was confronted with the problem of deluding three of the canny trout of the Letort, which were feeding in deliberate style on pale-watery duns beneath the spreading branches of several mulberry trees which overhung the stream. I recall clearly that I did not consciously select that pattern from my fly box but merely reached into the compartment of pale wateries in automatic fashion to obtain a fresh replacement for the sodden fly on my leader. After a moment of hesitation, I knotted it to the leader and threw it to the lowest of the three untouchables. The float was perfect and there ensued a deep gulping noise, unfathomable to me for a moment; then the glowing realization of a successful deception prompted the strike and the subsequent play and capture of the trout.

The lower of the remaining two was handled in the same fashion, requiring two floats to raise him. The third and last succumbed to the same blandishments on the first float, and my confusion was complete. Here were three of the Letort's fine trout, heavy and deep-girthed, who had taken this questionable item in that deep fashion which I like to believe is a certain sign of a successful fly. I could not understand it. I will confess that I never used the fly again, keeping it secure to look at often, and my wonderment has grown with the passage of time.

I am certain now that it can be explained on the basis that it had a good light pattern, not distorted by a long, submerged body and tail, and a fairly good representation of the pale-watery wing. But it needed a number of improvements, chiefly a better balance, for without tail support it had a dastardly habit of falling flat on its backside, thereby altering the light pattern and killing the trout's view of the wing. The fault could not be rectified by pulling it upright without the fatal consequence of drag.

What is more rational, more practical, and better conceived to overcome this difficulty than shifting the position of the hackle toward the bend of the hook, obtaining maximum support with stability closer to the center of balance by using two short-fibred hackles turned at opposite angles in the manner of an X? The result is that the afterbody is shortened (which is entirely correct, since the majority of duns carry the afterpart almost perpendicular to the surface of the water, hence presenting a foreshortened view to the trout) and the tails are elevated skyward, exactly as they are carried by the living dun.

THORAX

Far more important to the trout is the presence of the forebody or thorax. Without exception, this part of the body hugs the surface of the water very closely, oftentimes touching it. It is sometimes part of the light pattern beyond the circumference of the window and is most prominent in the window itself. As with the submerged body of the spinner which has broken through the mirror, it is startlingly clear in outline, and possibly in color, to the trout. In any event it is a happy circumstance that it exists for the flytier to imitate, since it offers a means of counterbalancing the weighty bend of the hook. Finally, the presence of the leader, attached with the Turle knot on down-eyed hooks, influences the head of the fly, tipping the balance forward, and provides a stiffening factor, permitting the head of the fly to go no higher or lower than the level of the leader itself.

TURLE KNOT

There is nothing novel in moving the hackle toward the bend of the hook. The palmers exemplify this prin-

77

ciple but disturb too much of the surface and exaggerate the light pattern for both duns and spinners. In some instances the construction of the fly has been completely reversed by placing hackle and wings at the bend of the hook, but these flies were never satisfactory in performance or appearance, and were undesirable anyway because the leader interfered with the tails, and because tying to the eye was troublesome. Even where the hackle was moved backward in the conventional tie, as was advocated by Dr. Burke of the Anglers' Club of New York, no thorax was added to complete the outline and the hackle was not arranged to eliminate the tail support. It was a worthy improvement in fly dressing, but it was incomplete.

BURKE'S FLY

Naturally, if the thorax is placed in front of the wing, as it should be, the wing must move backward with the hackle in order to maintain the correct proportions of the natural. The wing of the living dun is anchored well forward on the shoulders, but the width of the wing behind the anchorage point extends the rear edge beyond the center of the body toward the tails.

Moving the wing to its proper place is not enough. If it must be given the full value that it deserves, something more than the conventional treatment is necessary. The standard tie for wings, with any of the usual materials, is patently defective in a number of ways. The starling and duck feathers are fragile things, depending for cohesion upon interlocking fibres, which are quickly rent apart by a few casts. Wood-duck and hackle-point wings make no pretense of maintaining shape and have the annoying property of becoming thin and wispy when they are wetted. They are sadly lacking in the one great necessity, namely, width of wing as it exists in the living dun and spinner.

That is the prime reason for the use of cut and shaped

78

hackle wings in the new patterns of duns. Note, however, that they must be cut from the *webby* part of a broad neck hackle, taken as close to the point as possible in order to retain flexibility of the rib. If they are cut close to the base of the hackle, the stiffer part of the rib will enhance any inequality in the paired wings and the fly will spin or whirl. Most delightful of all, however, is the fact that the correctly cut wings do not become split or disarrayed because of use or abuse. The fibres are stiff and independent of one another, yet yield with the bending of the flexible rib. They maintain their shape consistently, and even when thoroughly sodden they are easily dried by a few extra false casts, which permit them to regain their former symmetry by self-recovery of the springy fibres. It is a quality which must be emphasized and secured at all costs, for *the wing, its height and breadth and flatness, is the most important part of a floating dun!*

It is the part that a trout sees first, for it must be remembered that the wing is visible before the body of the insect comes into the window. The light rays entering the window at the lowest angle, almost parallel to the surface of the water, are intercepted by the upright wing beyond the circumference of the window; then they are bent sharply downward, causing the image of the wing to appear above the window, tilted forward and sliding downhill toward the trout. Of course, it quickly rights itself in the view of the trout as it approaches the center of the window, since the rays of light are less and less refracted as they enter nearer to this point. And not only does a trout get a very good view of the wings, but he sees them better by virtue of varying contrasts as they move along before the landscape.

These are good reasons for the success of the one-fly man at least 50 per cent of the time when fishing to duns, if his one fly in different sizes has a blue or gray dun wing, that dusky indistinct hue which is characteristic of so many duns; he might be even more successful if he had one more pattern with a pale or yellowish-white wing to accord with another large group.

These are the two main divisions of the duns and they should be classified accordingly. The important thing is not so much the difference in color as it is the great variance in light and dark between the two classes, the one being subdued and absorbing much of the light, the other reflecting and flashing so much light that a trout could not mistake it for the first. It is a distinction which has no entomological significance, but that is no loss for those of us who must deal with optical facts.

To be sure, a trout does not eat wings. He eats bodies, with which he has a lifelong familiarity both above and below the surface. If the wings did not have a body attached to them, tactile experience would soon teach him that they were not worth eating; but he cannot know this until he has first taken the wing, for it is extremely doubtful that he ever sees the body, except perhaps for the thorax, and that only when it is in the center of the window against the perpendicular rays in that area. With the imitation it does not matter anyway, since it is necessary to fool him only once.

There are physiological as well as geometrical reasons for his inability to see or, shall we say, notice bodies. A flat plane with an unbroken surface will catch and hold the eye to the exclusion of other parts which do not have the same light values. The plane represents the area of greatest reflection of light rays and the eye, whether of trout or of man, is strongly inclined to focus on this area. All objects, animate or inanimate, create a visual impression involving parts which stand out in bold relief, other parts which disappear into umbra, and still others

80

which recede into penumbra. It is impossible to see all of the features simultaneously. As applied to the floating dun, it follows that even if the body is visible it makes no appreciable impression on a trout whose attentions are riveted and held by the comparatively wider, flatter expanse of the brilliantly lighted wing.*

* In Chapter X of Halford's autobiography he tells a remarkable story of how, in 1886, he and his good friend Marryat were fishing together during the green drake season and had their lodging at Houghton Mill on the Test. After a very hot and exhausting day astream, both of them returned to the mill in the evening and retired early to sleep soundly until the following morning, when Halford was roughly aroused at 5 A.M. by a pajama-clad and excited Marryat who was alarmed over the fact that neither of them had a decent supply of artificial mayflies for the day's fishing, having lost or given away nearly all of their stock on previous days.

Urged and prodded by Marryat, Halford was forced to get up and busy himself with vise and materials. He was, needless to say, in a somewhat sleepy and disorganized state of mind, and cross too, though he found it extremely difficult to show temper in the face of Marryat's usual high good humor.

The task of dressing each pattern was divided between them; one to attach the wings and the other to fashion body and tails and turn the hackle. Halford was occupied with the latter task, but so great was his disorganization that he astounded both himself and Marryat by completing the patterns without remembering to put on the bodies, the finished flies having only tails (or whisks), hackles, and wings. Marryat was quite intrigued with this accidental result and rather admired it, and he promptly dubbed it the "ghost."

All together, two dozen flies were dressed on that early morning session and half of them had no bodies. What is most remarkable of all, however, is Halford's positive statement that subsequent experience showed that the "ghosts" were as successful as the flies with bodies. Both of them used the ghost patterns for a time thereafter with complete satisfaction and then abandoned them forever because it *offended their taste.*

What a strange conclusion for such a wonderful discovery! It is likely that two more skilled or more observant dry-fly anglers than Halford and Marryat never lived, yet neither of them caught the significance of this important revelation.

It is often said that there is nothing new under the sun, and that is undoubtedly true as far as the physical evidence of a fact is

HALFORD'S GHOST

With the spinners or spent imagos and some of the terrestrials such as the grasshopper or Japanese beetle it is an entirely different matter, since their bodies are visible in a direct line on the underside of the mirror long before they have entered the window; and besides, the wings of the grasshopper are transparent and watery in appearance while those of the Japanese beetle are folded and integrated with the back, hence invisible. It is plain that in these cases it is imperative to construct and fish *a body* only.

How different from this, for example, is the Halford or standard tie for the ant, a terrestrial which usually lies sodden and partly submerged! The Halford tie is indistinguishable from the form required for floating duns; the wings are tied upright and are of opaque material, whereas those of the natural are watery and lie flat in the surface film as do the wings of the Ephemera spinner. And the two furnace hackles used to imitate the legs are completely unreasonable, obscuring as they do all

HALFORD'S ANT

concerned; but there is always something new and exciting about the discovery of the true meaning of that fact. Both of these men seemed to have been bound too narrowly by a conservative attitude which prevented them from venturing into the unknown field of the trout's point of view, else they might have established for all time the significance of wings in the trout's field of vision; or perhaps it was more correctly the result of strict adherence to the classical concepts of that era, which demanded realism in expressing the particulars of a work of art. This same attitude in some of the modern writers appears to have sapped the effectiveness of their works. Col. Harding's splendid treatise, for example, contains a wealth of factual information but seems to be trammeled by a fear of advancing new conclusions; it seeks in almost every case to explain and justify traditional or classical practices when a different and more reasonable interpretation could have been drawn from the facts.

G. E. M. Skues, a contemporary of Halford, might have accomplished the emancipation of the dry fly if, instead of a few "fugitive papers," he had performed a formal and complete work similar to his daring treatment of the wet fly.

of the body value which is paramount for this insect. What's more, the only concession which is made for the comparatively huge afterbody of the ant in the standard tie is a small ruff of unstripped condor quill at the bend of the hook. It violates all of the sound principles of light pattern and form for a given insect.

Apropos of these remarks, there is an interesting and absorbing account in Chapter VII of Romilly Fedden's *Golden Days,* one of the most delightful books on the entertaining side of angling ever written.

The account concerns a giant trout feeding industriously on natural Alders below a bridge. While Fedden watched from the bridge, a companion was busily engaged in the business of trying to lure the big fellow with a standard Alder pattern tied with upright wings. He was making each cast as perfectly as possible, but each time the fly passed over him that fish rolled up, looked once, and returned to his original position.

In the meantime, Fedden amused himself on the bridge by manufacturing admittedly clumsy imitations of the Alder with short lengths of dried grass, strips of rubber, and Harris tweed, but with wings fastened flat and not upright. These were dropped over the parapet to the waiting fish, who accepted them confidently one after another. Fedden was joined on the bridge by his friend, who argued that the fish had probably changed his tastes and would in all likelihood take the standard pattern of the Alder if presented in the same manner.

One of these was taken from a fly box and thrown to the big trout with its barb removed. The fish rose, looked, and refused it. Then Fedden dropped an Alder of his immediate manufacture and it was taken. This game was played for some time, fly for fly, and in the end the score was, as Fedden puts it "six-love" in his favor.

It was then suggested that the fish be tried again with

a standard Alder, manipulated so as to resemble the natural by tying the wings flat and clipping away all hackle from the back. The fly was attached to the leader, the cast was made, and the great trout fastened to the "doctored" Alder without hesitation. The ensuing contest resulted in a break and the loss of the fish. Other incidents similar to this have been recorded in books, but none of them is more complete in detail than this one, in spite of its casual nature. That the absence of an upright wing in the dressing of the Alder was a deciding factor seems to be a fair conclusion.

ALDER

Out of all this, there is a strong suspicion that trout fasten their attentions upon a single outstanding feature of each separate class of insects, and that it is this feature which causes them to rise or not to rise. It may be the body in one class of insects or the wings in another, but only one thing at a time. If something is added which overpowers and draws the trout's attention from the essential similarity with the naturals, he is likely to hesitate, not because he is influenced by differences but because he has not seen the similarity. He must be given a chance to exercise his limited perspective.

So far as the fulfillment of the prescription for a proper dry fly is concerned, it is comparatively easy to accomplish, but serious difficulty may be anticipated in satisfying the requirements for color and translucency in the essential parts.

Long ago, as far back as Charles Cotton's time, the fact of translucency was noted and appreciated by fly fishermen. In Chapter VI of the "Second Day, Part II" of *The Compleat Angler* the dialogue on fly dressing is in the following words:

84

"PISCATOR: . . . so, here's your dubbing now.

"VIATOR: This dubbing is very black.

"PISCATOR: It appears so in the hand: but step to the door and hold it up betwixt your eye and the sun and it will appear a shining red; Let me tell you, never a man in England can discern the true color of a dubbing any way but that; and therefore choose always to make your flies on such a bright sunshine day as this. . . ."

I think that these instructions are as sound today as they were in the 1600's, when fly-dressing was limited in scope, and Cotton's renown must be enlarged by observations as clever as this one. But this, the only expression on translucency by Cotton, did not contain all of the discoverable elements on this subject, nor does it even appear that he related it to specific insects.

Halford made some provision for it in the bodies and wings of a few spinners, by the use of horsehair on the bare hook shank and hackle points for wings. Later, J. C. Mottram offered some pertinent suggestions in his *Fly-Fishing: Some New Arts and Mysteries,* but it remained for J. W. Dunne, in *Sunshine and the Dry Fly,* to open wide the doors to the citadel of translucency and color, heretofore only dimly perceived by the fly dresser of tradition. Whether or not he has achieved the desired effect in his "Sunshine Fly," he must be credited with a worthy effort to reproduce this elusive quality.

"Against the light" said J. W. Dunne and thereupon he pinned his entire faith in exact imitation on the factors of translucency and color.

"Against the light," he exhorts, when the living insect is being viewed for the purpose of imitation.

Now that is a better admonition than Cotton's, for it definitely relates translucency not only to fly-tying mate-

rials but also to the living insect itself. That is a far-reaching step, one worthy of emulation, for it is a fact that color in a living insect can be remarkably changed when viewed by transmitted light, i.e., when light passes through body and wings as they must appear to a trout under certain conditions. That these conditions exist there is no doubt, yet there seems to be a studied generalization on translucency in Dunne's book which is not warranted by observable facts about the living insects. A careful reading conveys the notion that all of the patterns are treated alike, not allowing for different degrees of translucency in different insects.

A close examination of many species viewed "against the light" will disclose some surprising things. Consider, for example, the startling fact that in two different species of the same family, the black and the red ant, there is a tremendous variance, the former being absolutely opaque in the body and the latter glimmering and glowing as though lighted by some inner fire. Even allowing for the high magnifying power of a trout's eye, which is easily duplicated with a magnifying glass, there is no apparent translucency in any degree in the black ant. Obviously these two must be dressed to obtain different effects.

Different species of the caddis family disclose a similar variation. The Japanese beetle and the grasshopper belong in the opaque category. So too the jassids. All of these except the caddis are terrestrial-born and -bred; and it may be argued that they do not invalidate Dunne's procedure for Ephemerids. Nevertheless, the same situation exists in the Ephemeridae; strange as it may seem, there is at least one of the mayfly family which has a spinner whose body is opaque. That is contrary to the generally accepted notion that all of the imagos are translucent. In order to make sure of this fact

86

I captured a great many of the duns of this species and permitted them to molt successfully in cages for a more thorough and repeated examination of the spinners. The result was always the same.

In the main, however, the spinners are highly translucent in the body and never more so than after oviposition, the final stage of their existence, when they are taken by the trout in quantity. There is another disturbing fact, too, for it is plain that the spinner becomes so transparent because it has shed a thick, darkening pellicle from the wings and body which it bore in the dun stage. The dun, then, is much less translucent than the spinner, and the thorax in both of them is entirely opaque. If, as explained before, the body of the dun is not observed by the trout, it is useless to reproduce translucency at this stage; and the same considerations apply for the wings.

How much, then, is translucency worth in the imitation of insects upon which trout feed? It is clear that in some cases there is none to imitate, in others there is not enough to warrant special treatment or materials, and in a few cases there is so much of it that it cannot be ignored. It is equally clear that all dry flies, even for the Ephemeridae, cannot be logically tied with the same materials (such as the synthetic floss which Dunne used for all of his patterns of the mayfly), for each insect must be separately rated and given its proper value in the scale of translucency. Every pattern must be described accordingly; this is not an impossible task, since the degree of translucency can be established, for practical purposes, by naming a well-known and easily obtained material which gives the desired effect in the artificial.

It must be carefully noted that translucency itself is not color. It only affects and changes color when viewed

87

against different backgrounds. Against a dark background the red ant has a dull, red-brown color, but when it is held between the eye and a bright light (not directly in line with a blinding light or the sun, for it is then a black silhouette) it becomes a light, golden-red color, the result of the filtering out of some of the color by the passage of light rays through the semitransparent body of the red ant.

Dunne obtained a similar effect by painting the hook shank with white enamel, covering it by a body made of a synthetic floss called Cellulite, and, finally, applying a special oil which thoroughly saturates the body and causes it to acquire a certain luminosity when viewed against the light. There is reason to suspect that this procedure was much more effective for the dressing of spinners than for the duns. It is a fly which is not generally known or used in America, partly because it is not usually sold in the tackle stores and partly because of the difficulty of obtaining the specialized materials. To my knowledge there was one group of ardent fly-fishermen who imported and systematically used the Dunne series of patterns on Pennsylvania waters for an entire season. The outcome of the experiment indicated that these patterns were no more successful than others. In my opinion it was an unfair test, however, since the imported flies were tied expressly for British waters to imitate British insects and since the general practice here of fishing the water instead of the rise is not a decisive trial for specific imitations.

Apart from the question of effectiveness, the complexity of following the tying technique is an objectionable feature to amateurs and professionals alike. Cellulite floss is an extremely sensitive material, inclined to separation of the fibres, which like to catch and cling to hangnails, fissures in the skin of the fingers, and ragged edges of

88

the nails. When it is wound on the hook shank there is a tendency of the floss to splay apart, causing some of the fibres to slacken and form unsightly loops. Then, too, it has a slippery nature which makes it impractical for the bodies of large flies if they are to be made as thick as the natural. The building of a large, tapered body often results in having the larger turns slipping downward over the smaller turns, and they cannot be recovered without creating unevenness in the floss and the afore-mentioned loops.

In tying imitations of the Ephemeridae, the fly dresser must bear in mind a very important point in connection with bodies, namely that our American Ephemeridae are usually wide and flat, a characteristic of the creeping and clambering nymphae. They differ enormously from the bulk of British duns that emerge from the swimming nymph, a round-bodied, slender creature. In either case the body of a winged dun or spinner is an exact counterpart of the nymph with which it is identified. Our American nymph fishermen are knowledgeable people who understand this principle thoroughly, with the result that they have fashioned a wide, flat artificial nymph with a prominent thorax, of wonderful attraction to the trout.

Except for a minimizing of the afterbody of the dun for reasons already given, that is the way in which to construct the bodies of many Ephemerid duns, with particular emphasis on the forebody. It calls for a lightweight, waterproof material to provide the requisite bulk; that is the reason for the choice of spun fur for some of the present series of dry flies, particularly the duns. It is a material which has many virtues and few faults. The catalogues of the suppliers describe it as a yarn made from the fur of rabbits, and it is generally found in most of the dry goods stores in any community. In the better grades

89

it is fine grained, smooth, and fluffy, without a multitude of fibres protruding in all directions as in the case of wool yarns and homespun fur. The smooth, waxy appearance of the duns is not better achieved by any other substance, particularly after the fur is wetted. The fringe of water clinging to such a body imparts a lifelike and sympathetic appearance to the natural.*

* Rabbit fur has long been recognized as a superior medium for representing the bodies of flies. There are many references to this material in the older books—all of them of an exclamatory and admiring character. At one time the Hare's Ear, a rabbit-fur pattern, was Halford's favorite fly; in his *Theory and Practice of Dry-Fly Fishing* he said that if he were limited to one pattern of fly he would unhesitatingly choose the Hare's Ear, a fly which had afforded him great success with the trout—and yet when his *Modern Development of the Dry Fly* was published, he had abandoned the Hare's Ear so completely that not a single pattern of this nature was included in his list. I believe that this same pattern was also the favorite of Hall, a contemporary of Halford's and the originator of eyed hooks. G. E. M. Skues expressed a similar enthusiasm for this pattern, and J. W. Dunne conceded that it was the only fly which compared favorably with his cellulite-body "Sunshine Fly" when viewed "against the light." In that comprehensive manual, *Trout Fishing From All Angles*, Eric Taverner ponders the merits of rabbit's fur and then bursts forth with the exclamation that he would give much that he possessed to know the secret of the fur's fatal attraction to the trout.

After the above lines were written, my good friend and fellow member of the Fly-Fishers' Club, "Bob" Bates honored me with the loan of his precious copy of Aldam's *Quaint Treatise on Flees and the Art A Artyfischall Flee Making*, a rare and glorious book containing thick pages with correctly tied artificial flies, and their materials, embedded in them. It was edited by Aldam and published in 1876, based upon a manuscript written about 1800 by an old man whose name remains unknown.

Imagine my unbounded delight when I discovered, among many other delightful things, that one of the Old Man's tremendous secrets was the employment of rabbit fur. The reference is contained at the very end of the manuscript, where he details the method of obtaining a yellow color for the body material. This section is fully quoted below for those who do not have the opportunity of seeing this rare book, and I have retained the language, spelling, and punctuation as it appears in the treatise to

The sole objection to spun fur lies in the assertion that it lacks a resiliency of fibre, thereby causing a matting or felting which detracts from its original excellence. The objection is more fancied than real, for although the fur has a matted appearance where it is dry, once it is wetted

---

help the reader understand why the word "quaint" describes the treatise so perfectly. Take note especially of the Old Man's sound reasoning and his deep understanding of fly-dressing problems. The words in parentheses are modern spellings of some of the more obscure words.

HOW TO MAKE YELLOW CARRITED STUFF

"Take the white part of Hare or Rabbits belley—then take one table spoonful of Aquafortis and tow (*two*) of water mixt them togeather—then by the acisstance of a ragg at the end of a short stick and a fark (*fork*) to keep your fingers from being bruned (*burned*)—lay the Hares belley upon a plate and with acisstance of the fark hould it fast and wett it well down to the roots with the mop—then hold it before the foir (*fire*) with the fark untill it is gone yallow—when yallow enough wash it well in could water to kill the Aquafortis—and when droy (*dry*) it is fit for use—this and a little blue Rabbit well mixt will be made to any shade suitable for all the Dun flees that is required in the Art a Arty-fichall flee making—It makes your Flee much nater (*neater*) and comes more to nature then that stiff brisley (*bristly*) Dubbing—You find nothing coace (*coarse*) in nature—when you have made a Artyfichall Flee as nate (*neat*) as hand can make it is a thousand times behind a natural one when dresst with the natest meatearills —When wee come to Examin thoes small beautyfull tender del-lagate (*delicate*) and nate water bred Duns that ought to be the Anglers coppiing—I can find no room for coace meatearills—the natest are very coace when compared."

All of this is very interesting and worthy of much speculation but there is real food for thought in the fact that the dry Hare's Ear of ancient fame was never tied with hackle; the body of rabbit fur was picked a little on the underside to give an impression of legs and it needed nothing else to make it float well—such is the buoyancy of rabbit fur. But besides this, there is a reasonable suspicion that this method of dressing (to my knowledge, the only pattern dressed in this way) permitted a small, neat light pattern, and more important still, that it allowed the wings to attain their full value in the eyes of the trout, not having been distorted or obscured by a mass of offending hackle.

91

again capillarity causes the body to swell slightly and the fine fibres are released. It is similar to the quality of a marabou winged wet fly, which looks like nothing out of the water but becomes in water like a thing possessed.

Let there be no alarm about the wetting of dry-fly bodies dressed with spun fur. With or without the aid of waterproofing agents they continue to float exceedingly well, requiring no more than the usual number of false casts to keep them sufficiently dry. All furs have the peculiar property of being able to imprison small air pockets which keep them buoyant, and, moreover, they do not change color when wet.

The theory of fly dressing for spinners has been mentioned only incidentally in the foregoing discussions, and rightly so, for it warrants a separate and specific handling, certainly more than the unconscionable act of tying the dun with spent wings to imitate the spinner. They cannot be identified so loosely. In reality the spinner is not a dry fly at all but a damp fly, as Col. Harding puts it, something which is halfway between a dry fly and a wet fly. In tying the spinner we are confronted by vast difficulties which make the tying of a dun seem comparatively easy.

When the dun emerges from the water in the form of a winged insect, his movements, from the point of issue until he finds a resting place in the streamside foliage, are laborious and clumsily executed—understandably so,  since he is burdened with an extra layer or envelope of chitin which entirely covers him, his wings, his body, and all of his appendages. When this brief interlude in his life span, lasting not more than a day or two, has passed, the dun bursts through this extra pellicle and emerges in the final form of its existence, the imago. A wondrous change takes place. Where before it was a dull-looking, ungraceful creature, it has become a thing of sparkling,

92

Fly Patterns

ORIGINATED AND TIED

by

VINCENT C. MARINARO

MAY FLIES

GREEN DRAKE DUN

GREEN DRAKE SPINNER

HENDRICKSON DUN

HENDRICKSON SPINNER

SULPHUR DUN

SULPHUR SPINNER

LIGHT OLIVE DUN

DARK OLIVE DUN

All of the spinners are tied half-spent. The smaller spinners vary from the text; their bodies are constructed of seal's fur rather than porcupine quill. When available, seal's fur affords a simpler technique for tying the smaller flies. Nothing surpasses the largest porcupine quill for floating the big Green Drake spinners and similar patterns. Early examples of porcupine-quill spinners revealed a weakness where the quill was tied to the hook shank as a result of a hinging action. This weakness was eliminated by inserting the point of another quill into the open end of the body quill and cutting the inserted point flush with the body before tying it to the hook shank.

The Large Black or Brown Ant represents another good application of the porcupine quill, but wrapped around the hook shank instead of being used as an extended body. The smaller ant is tied with translucent horsehair. This technique should be limited to the smallest sizes—20 to 28. Larger sizes do not float well. Hollowing the quill for the spinners provides maximum translucency, but this is optional because white or cream quills already possess a great degree of translucency. It is important that quill and horsehair be thoroughly soaked and washed in soapy water before use. A size-14 treble hook should be used for the Pontoon Hopper; two of the hooks should be cut off to form a firm cradle for the slippery quill.

All flies except one Jassid, which has been enlarged for detail, are actual size.

TERRESTRIALS

 BLACK OR BROWN ANT

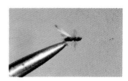 DARK RED ANT

 BEETLE OR DOUBLE JASSID

 JASSID .

JASSID (2:1)

PONTOON HOPPER (GRASSHOPPER)

diaphanous beauty, quick and sure in movement, light and airy as the fabric of a dream. Where before the body was sad and sober-looking, it is now charged with a soft, glowing color. The wings are of a glassy iridescence, receiving and giving a multitude of rapidly changing tints, externally defying the skill of ages in the attempt to achieve the sum total of these effects.

Here, indeed, is the translucency that J. W. Dunne extolled, and here is the severest challenge to the flytier's art that nature could devise.

Good or bad, the initial effort to imitate the imago must begin, as before, with the appearance of the natural on the upper side of the mirror beyond the area of the window, where the trout usually receives the first indication of the insect's presence. Yet, even before that, there are some things about the spinner that are worth knowing and worth applying to the imitation.

The general conception of the spinner's position on the water is one of body, wings, and tail lying fully extended in the surface film of the water, as though the spent female had suddenly collapsed in mid-air and fallen flat to the surface of the stream. That is only partly true, for many times the first contact with the water results in an attitude which closely approximates the position of the dun, that is, with wings upright. Spinners may ride the water for a surprising distance in this manner; but then their wings begin to droop, gradually falling away from the perpendicular on either side of the thorax until they finally become flush with the surface and all other parts are lying equally prone.

On the other hand, there are the spinners of some species which do not become fully prone but fall over on one side or the other, one wing lying in the surface film and the other in the air with stylets flaring wide and high, free of any contact with the water. This is con-

sistently true of those spinners that carry a ball of eggs in the little hollow at the end of the body which is curved downward and forward while the insect is in flight. When the egg mass is released in the air, the afterbody never again resumes completely the upcurved position of the dun but recovers sufficiently to form a straight line with the thorax, with the stylets remaining locked in a perpendicular position.

These are characteristics which affect the light pattern with which we must begin an accurate representation of the spinner. If this is true, it follows that the traditional prone position of the standard tie is not a proper execution of all the imagos. The prescription must vary to fit the individual species.

What is most surprising of all, however, is the transformation which takes place in the afterbody of the spinner when the egg mass has been voided. Look to the spinner after oviposition, when the trout get it, and it will be clear that the fine color which the body once had was largely supplied by the color of the eggs; nothing remains but a dull semitranslucent tube of chitin. It is astonishing to note, too, that so many different species have the same appearance after oviposition. Thus the female spinners of the Green Drake, the blue- and pale-winged Sulphur, and the light Cahill are so much alike that they can be imitated with the same body and vary only in size. They are uniformly of a chalky, whitish color.

In another group, the remaining color is a brownish cast varying from tan to a dull red-brown. These are the sole requirements of any importance for the spinners; I do not consider the variations in males as worth mentioning since they are seldom on the water. I except only the male of the Green Drake, but more of that later.

What is needed, then, is a medium which best represents the hollow, chitinous, semitranslucent body within the limited shades prescribed above.

In casting about for a solution of this problem I do not recall that any of the known procedures in tying spinners were overlooked, including the Halford technique of horsehair on the bare hook shank and J. W. Dunne's paint—Cellulite—oil covering. I cannot quite bring myself to the point of omitting translucent parts, as J. C. Mottram suggested. Trout are such unimaginative creatures! Besides, no matter what degree of translucency exists in the insect, even though it be clear as window glass, the substance still makes indentations of the water which cannot be omitted, in other words, a light pattern.

A satisfactory solution came about in a curious way. During the course of a meeting with the estimable Bill Bennett of the Harrisburg Fly-Fishers' Club, he chanced to mention the use of porcupine quill in preference to peacock or condor for some of the standard patterns, and forthwith gave me some to try. He explained that it was a comparatively recent discovery and was applied in the same manner as other quills, that is, by tying at the bend of the hook and winding around the shank to the shoulder. I was immensely intrigued by the waxy white color, the cylindrical shape, and the ease with which it could be manipulated; but when, at the next opportunity, I tried to use this material for the bodies of spinners, I was disappointed to find that the quills, on being turned, became flat and thin, losing their cylindrical shape entirely. In addition to this the thin bodies did not show a vestige of translucency. The remainder of the quills were carelessly thrown aside on the fly-dressing table, where they remained for some time in full view and unnoticed.

A few weeks after this occurrence, I was idly plucking and examining bits of odds and ends on the table without any distinct purpose in mind. The porcupine quills caught my attention and I can recall a fleeting moment of regret that such an interesting and suitable material as this could not be used for the bodies of spinners. Then suddenly a thought occurred which prompted me to clear the table to try a novel, experimental tie with the porcupine quill.

Briefly, it involved insertion of the hook shank lengthwise through the quill to form a body. The result was unsatisfactory in many ways not necessary to enumerate. I had hardly completed this abortive attempt when the answer became immediately clear. I fastened a clean hook in the vise and chose a new round quill from one end of which I cut a short piece no longer than the afterbody of a medium-sized spinner, about ⅜ inch. I then anchored the short piece at the bend of the hook by tying down the cut end to form an extended body, parallel with the shank. An immediate inspection against the light disclosed a body that was not translucent enough.

Beginning anew, I prepared another quill by inserting a twist drill of suitable size and with a few turns quickly removing the soft, pithy interior. The quill was attached to the hook as before, then held against the light and lo! there was the body for which I had searched so long— the chalky, chitinous, translucent tube of the spinners!

The rest of the tie was easy. Spun fur for the thorax, since this part of a spinner's anatomy is only faintly translucent or nearly opaque, and hackle fibres for wings, arranged in a novel manner to be described in a subsequent chapter.

Take note above all that the pattern calls for a hook with short shank which is completely enclosed in the opaque part that corresponds to the thorax; but the liability of the hook shank in the essential or translucent

96

part, over which J. W. Dunne labored so painfully, has, at last, been completely eliminated!

The extended body described here is not to be confused with the objectionable and upcurved extension affected by Louis Rhead. It has no curve in any direction, and it is no longer than the afterbody of the spinner. It has no opaque foundation such as gut, boar bristle, or hook shank, and it has no multifarious wrappings of silk, ribbings, and the like.

Experience with the porcupine quills revealed their many virtues one by one. It is the only quill-like substance which yields to the pressure of the finest tying silk, allowing it to be bound securely to the hook, yet possesses a remarkable durability. It is absolutely waterproof, since the open end is bound so tightly to the hook that water cannot enter. At the root end of the quill there is a small tag with a small knob, perfectly designed to hold tail fibres bound on with a few turns of silk and a drop of varnish. The next glad surprise occurs at the other end of the quill, where can be found the red-brown color required for another group of spinners in the brown-bodied class.

To establish a more complete understanding of the quills in question, I must explain that these are the short, slender quills found underneath the porcupine's body hair; reasonable caution should be exercised in handling them lest the fingers be pierced by the sharp end and inflict a painful injury, a property for which they are notoriously known. It is a good plan to blunt them by cutting off a very small bit of the point, resealing later with a little varnish when the tails are attached.

So much for the structural necessities and translucency in dry flies. There remains only the question of color, a subject upon which there is great diversity of opinion and unjustifiable controversy growing out of ill-considered opinions and superficial observations. A review of the definitive work of scientists on the color perception of fish ought to be the only sound basis and beginning for discussions of this, the most obscure and often disputed element in any formula for flies, dry or wet.

To treat the matter of color in artificial flies in any other way is totally inadequate and unfair to the reader. We cannot even be guided by the artist, whose training and appreciation of color does not necessarily coincide with the views of a trout. There was a time when flytiers were untroubled by the findings of scientific men, when color was accepted as a phenomenon of equal value to all of the seeing animals on earth. It is an attitude which is plainly reflected in the Halford and pre-Halford literature and sometimes in modern angling works.

The limitations of the flytier in the matter of reproducing color are measured, concurrently with the growing and appalling fund of knowledge adduced by modern investigators. These findings for the most part are not conclusive in themselves but suggest further investigation, leading the researcher deeper and deeper into a maze of inquiry that seems to have no end.

For example, it does not make the flytier happier to know that at last science has proven that fish can perceive color and distinguish fine gradations of color. There is the additional discovery that fish can see and appreciate color in the ultraviolet field—color in wave lengths too short to register on the human retina, colors we have never seen and cannot identify.

It does no good to say that if a trout is feeding on yellow mayflies, give him a yellow artificial anyway and let him translate it in whatever terms he pleases, then we shall still be correct. You can never be sure, without positive scientific corroboration, that the pigments in the yellow artificial reflect colors in the same range of wavelengths as the trout can see. The color which looks like yellow to the human eye may be something entirely different, not even a shade of yellow, if seen in a wavelength of 300 units of measurement, let us say, bearing in mind that humans cannot see below the measured limits of 400.

It is a problem which belongs strictly to the physical scientist and will finally be solved by him. At long last he will discover the limits of a trout's eyesight, the exact group of rays in which he sees an insect, how much of these are reflected by the individual insects, and what materials and pigments fit the formula for each of the naturals.

When these things are determined, it is possible that a mechanical device analogous to the photographer's light meter can be invented and made available to the flytier to assure him accuracy of color, not only in the range of the visible spectrum, but also in the fields at either end, the infrared and ultraviolet. That is the kind of accuracy that Halford and Dunne really wanted, and there is no other way to obtain it; for even if the colors in the invisible fields became visible to humans, the eye could not interpret them correctly, that organ having the faculty of automatically adjusting itself to various light intensities, rendering color and light values by no means the same as those perceived by a trout.

Until the time arrives when these things can be determined with the requisite exactitude, the fly-fisherman and flytier must be content with an attempt to approximate the color that he sees, not overly burdening himself with a meticulous and complex process of color evaluation. It means that he must continue to resort to the trial-and-error method until he discovers materials and colors which afford a reasonable amount of success. That method does not please those who preach a brand of purism in fly dressing which is simply unattainable at the present time, however desirable it may be.

There is a further limitation in the sobering and distasteful fact, easily observed, that there is no true uniformity of color in a given hatch of one species of mayfly. Let anyone try to establish for himself the correct color of the natural by examining and comparing not one, but many of the individuals of a single species during the hatch, and he will discover a variation so marked, in some cases, that he will forever despair of coming to a definite conclusion.

To complicate matters a little further let the examination be made by a group of observers at the same time, and if they are blessed with a sense of humor the outcome will be a most laughable experience. One of them will swear that the Hendrickson has a dirty yellow body, and he will be right, for some of them do have that color. Another, equally sincere, will swear as violently that it has a reddish-tan body; and he will also be right, for some of them have that color too. Yet a third, convinced that the first two are victims of a hitherto unsuspected myopia, will argue that the Hendrickson has a liverish cast. Finally they will go their separate ways, each of them marveling inwardly that the others should diverge so crassly from the plain truth of what he had seen so clearly.

Do not suppose that the Hendrickson is an extreme case of nonuniformity in color—not by any means. When my interest was centered upon the jassid, a small terrestrial of considerable dietary importance to the trout, repeated examinations made it increasingly clear that the great variety in species and color made standardization utterly impossible.

In order to satisfy my curiosity about the extent of this insect's variety, I requested permission, which was

100

granted through the kind office of Dr. Champlain, in the Bureau of Entomology of the Pennsylvania State Department of Agriculture, to examine the collection of jassids in that institution. Case after case of this insect was drawn from the cabinets for my inspection, each succeeding case exhibiting a bewildering display of color the like of which my streamside inspection had not prepared me to expect, careful though it had been. Every color conceivable was there, including many shades that I had never seen before, and I determined then and there to quit any more senseless attempts to establish a color standard, at least for the jassid. I trust that the reader will be only mildly shocked when he discovers that the tying instructions for the jassid contain the direction "any color."

JASSID

In the light of these and other revelations of the same character, it is impossible to understand the excessive demands placed upon the fly dresser and the fisherman by the exact-shade-of-color theorists.

Experiences indicate strongly that trout react similarly to many shades of a given color, where the patterns are tied with the same material colored with the same dye. So much seems to be confirmed by the experiments of Reighard and a few others; and if the reader wishes to pursue the subject further in all of its ramifications, there is no better beginning than Harlan Major's summary on the color reaction of a fish and the bibliography contained in his excellent book *Salt Water Fishing Tackle,* from which the reader is bound to emerge somewhat uncertain and considerably shaken in his preconceived notions about the color sense of fish.

Even so, there are some glaring absurdities in the arrangement of colors for the standard flies which ought to be avoided if color has any value at all. They follow upon a popular and perpetuated misconception about the

101

true function of hackle for dry flies. Consider, for example, the utterly unreasonable custom of imitating the color of legs with a similar color of hackle.

When Halford, or any of the modern flytiers who follow his plan, prepares to tie a certain fly, let us say an Iron Blue Dun, he fashions a wing of the proper slate-blue color and body and tails of the correct shade, then proceeds to dye or select hackle for the legs, which are supposed to be a red color. The hackle is tied in and turned perhaps five, six, eight, or ten turns (Halford customarily used two and three hackles) and what is the result? There is a mass of hackle which completely hides and obscures the wing, destroying the sharp, clean outline of that part; and what is more, the color of the hackle becomes the outstanding or dominant color, becoming the wing itself with an entirely different color than the original wing which the hackle displaces. What we really have is an Iron Red Dun, not an Iron Blue!

Small wonder that we often hear the angler complain that the standard imitation was a failure during the hatch, and that trout were caught when something different was used. Small wonder that hardly a vestige of faith remains in the theory of specific imitation. These instances are multiplied many times, as an examination of any catalog of descriptions will demonstrate.

The only sensible conclusion, then, is to employ hackle in such a manner that it *supplements* and *confirms* the color value of the wing, and that means that wing and hackle should be of the same color. A departure from this principle is justified only when a fly of small size can be tied on the lightest of hooks, requiring only two or three turns of short-fibred hackle to support it. Or, in other cases where the wing is not paramount, such as in that of the Japanese beetle, there can be no harm in imitating the color of the legs.

102

I do not discount the practicality of using a wing formed by hackle only, tied buzz fashion. If the turns of hackle are closely packed and the fibres are perpendicular to the hook shank on all sides to form a flat plane, it makes a very good wing, although it lacks stability without tail support.

If such a wing is used, by all means, the color of the hackle should imitate wings, not legs.

Lastly, there is the usage of hackle fibres to represent the setae of dun or spinner in such large number that they appear to be an extension or continuation of the body, actually being thicker in the mass than a hook shank thinly covered with quill or a similar material. The result is not only a distortion of body length but a suffixing of color differing from and devaluing that which precedes it. It is a common error in many of our American examples of the flytiers art, and it ought to be severely condemned.

How admirable is the modern English version of this unit, never requiring more than four fibres and oftentimes less, even for the conventional tie, which demands tail support!

The only excuse for the use of more rests upon the employment of outrageous hook sizes—12, 10, 8, even 6—too heavy to be supported by any reasonable number of hackle fibres. There is no fly on the American continent, not even the Green Drake, that needs anything larger than a size 14 for imitation, and there is no trout that rises to a dry fly which cannot be hooked and held by such a hook, with the proper skill which ought to be the property and pleasure of every fly-fisherman.

What, above all, is the sense of wagging our heads in resignation and saying that we cannot hope to equal nature's handiwork, of deliberately exaggerating our difficulties as though it were an act of virtue!

Note

THE SPECIFIC and detailed tying instructions for dry flies, according to the principles announced in the preceding chapter, will be postponed until last so that a fuller understanding may be gained by a reading of the following chapters, which are largely an account of experiences in using such flies to imitate specific insects. The insects themselves, with the possible exception of the Letort olives, are typical of those which exist everywhere in trout waters in the United States and Canada. The Hendrickson, I believe, was first named in the Northeast, on the Beaverkill, a typical freestone stream, and the most complete scientific study of this same insect was made by an eminent Canadian entomologist.

The blue- and pale-winged sulphurs are equally widespread, and even if species may differ a little, they are exemplary types of a pale, watery class of mayflies generally pale yellow in coloration.

The Green Drake is known to almost everyone and needs no particular introduction. It thrives on every kind of stream imaginable and on many lakes, too.

The terrestrial insects are in a class by themselves, and we should speak of their presence on the water as a

104

"flight," not a "hatch," since the emergence point or metamorphosis occurs on land and their presence on the water is purely accidental, as we shall see; but they are, nevertheless, quite as important as the aristocratic mayflies. The grasshopper in this group represents somewhat of a departure from the placid fly which dominates these pages. It is really an active fly, but I would not exclude it because of that. It is one of those odd cases where the principles of one method of fishing overlap those of another, but let us make no mistake about this; it is a perfectly legitimate form of fly-fishing, and I would not limit myself by abandoning it in the manner of Theodore Gordon, who would have nothing to do with any imitation except that of the mayflies. That is too narrow a view, particularly when we contemplate the enormous size of some of the trout that can be taken in grasshopper time. This insect and the ants are as well known as any could be. Indeed, I believe that the grasshopper is even more plentiful in the West than in the East, and it ought to afford spectacular fishing on some of the great Western meadow streams like the Yellowstone, Owens, and Gallitin Rivers and many others, particularly during the great flights which "darken the sun," according to the reports one often hears.

The Japanese beetle is the only insect of the terrestrial class the discussion of which might be of questionable value to those outside the area of its infestation; but that is something which may change with the passage of time unless some means to check its progress is discovered. In any event, it may be regarded as a beetle type and the imitation is marvelously good for representing any of the beetle family in a dry pattern, which ought to be light in weight and ovoid in shape.

The class of minute insects is distinctive in size only, for it includes the tiny mayflies as well as terrestrials; the

105

great problem here is the tying of these small artificials on No. 20 and 22 hooks. Most of us like to avoid them if possible but we do so at the great expense of losing the most fascinating kind of dry-fly fishing in the world.

Chapter 4

HENDRICKSON

W HILE the mountain torrents are still raging with the burden of melting snows, while the highlands are still sere and stark from wintry blasts, spring has already arrived in the beautiful Cumberland Valley, heralded by the yellow bloom of forsythia, first sign of the changing season. For the fly-fishermen of this valley there is no gradual transition from the tedium of winter to the gentler season, no measured progress marked by alternate days of cold and warmth; simply and suddenly there is the forsythia, faithful precursor no matter what the state of the weather, and suddenly there are thoughts of the Yellow Breeches and—the Hendrickson. The pursuit of the hatches has begun.

Following the time-honored pattern, there are likely to be a few excursions along favorite stretches of the

107

stream, little forays in advance of the opening day, to scan the water anxiously, hoping and yet not hoping (the hatch must not be premature) for a sight of the big slate-blue sail of Ephemerella Subvaria hoisted aloft on the placid current. If the sails can be seen, their numbers are roughly estimated; then follow some swift mental calculations, comparisons with number, frequency, and emergence dates of former years, perhaps a hasty reference to a pocket diary, all of which help to determine whether or not the main hatch will begin properly at or near the lawful opening of the trout season; for this angler is justified in expecting the finest kind of dry-fly fishing on the first day astream.

But nothing is more dismaying than to see great clouds of the spinners moving upstream in the evening in advance of opening day; for this means that the hatch has been in progress and that the peak of emergence will have passed before the fly-fishing begins. But granting that this has not happened, there are still some uneasy days ahead—days when the stream must continue to remain clear and unsullied. Even though this level valley is not noticeably influenced by melting snows, oftentimes the torrential rains of spring form rivulets bearing the mud and debris of fields and roadsides. Nevertheless, the records show that these hazards have not happened too often to mar the angler's anticipation, the majority of past seasons having been timely in point of hatch and water conditions. The slight variation in emergence dates would not matter very much in midseason, but a difference of two weeks, moving the emergence period ahead of the opening day, is enough to cancel out all of the fly-fishing to the Hendrickson. The angler must then content himself with fishing to the less satisfactory caddis flies, then in abundance, or, like his brother angler of the

108

North country, ponder the merits of wet flies, bucktails, streamers, and perhaps some less aesthetic devices.

Let us assume the existence of the happier circumstances for the dry fly. Then there is no better beginning for the fly-fishermen's year than to fish the Yellow Breeches with the Hendrickson, dun and spinner. Let us assume further that we shall encounter a hatch such as we saw three years ago when, following the appearance of a few stragglers, the hatch broke into full swing on the morning of April 15 all over the Breeches, with little flotillas of duns coursing along on every line of drift and being dispersed or decimated by the hungry trout with wonderful enthusiasm, to continue in this fashion, almost every day, evening too, for three exciting weeks.

Ephemerella Subvaria emerges quickly and easily from the nymphal shuck, appearing unexpectedly on the surface, performing the magical transformation of nymph to dun in that startling manner that is reminiscent of those familiar deceptions practiced by artists in sleight of hand. He rides for a great distance on the water, making comparatively little commotion, before taking to the air, owing, no doubt, to the inclement weather which obtains at this season of the year. There may be an intermittent, spasmodic opening and closing of the wings but nothing more than this. He is a most imperturbable mayfly.

On certain mild, sunshiny days he is inclined to shorten his rides somewhat, although he does not exhibit nearly the amount of activity displayed by the pale wateries, for example; and when he leaves the surface of the water he does so abruptly and efficiently. These are likely reasons for the sedate conduct of the trout, who take this insect deeply and with an assurance not surpassed by their attentions to other species of mayfly, creating in

109

almost every case a rise form with heavy concentric rings, accompanied by resonant, low-scaled sound.

On certain blustery days, the high winds (kite-flying winds) are responsible for behavior that is not usual with mayflies later in the season. Successive gusts will beat against the broad expanse of the upright wings, causing the insects to veer away from the regular line of drift, rocking along in a tipsy manner and careening across the stream like little sailboats tacking before a high breeze, with the result that nearly all of them can be found parading downstream in narrow file in the lee of the far or near bank. The knowing angler must adjust his approach and position accordingly. Not even the Green Drake, large as he is, behaves so entertainingly, but it must be remembered that Ephemera Guttulata is a weighty fellow, not easily influenced by breezy weather, and that he arrives in the balmiest season of the year.

The long drift of the Hendrickson and the narrow feeding lane assumed by the trout once they have got over the excitement of the earlier caddis flies makes it imperative to cast the artificial accurately and float it for a great distance. Trout always seem to adjust their feeding pace to the tempo and behavior of the particular insect which is on the water, and the feeding on the Hendrickson is no exception. Nothing is more conducive to lack of success than to keep hammering at the rise form itself with quickly repeated casts, floated short, in utter disregard of the deliberate and studied pace of the trout. But more than this, trout exhibit a certain amount of rhythm in their feeding, never better demonstrated than during a heavy, sustained hatch of the Hendrickson.

In the early stages, when they are hungry and eager and the naturals are coming down in little fleets, they may range a little wider than usual, taking those insects which float along a little to one side or the other of the

110

strongest line of drift. At this stage the angler must be careful not to cast right-face when the trout is feeding left-face, and it may be necessary to search across the line of drift with a few casts in order to locate and attract the ranging fish. As the hatch and the feeding progresses the trout will narrow the groove of his feeding lane considerably, and the fly must be cast and floated more accurately than before. When the point of satiation is approached, the trout will likely accept every third or fourth insect or, seemingly, only those which coincide in point of time and place with his lagging desire to rise again. In any event, it is necessary to guess correctly when he is ready to take another, and to cast the artificial into the breach before a natural is taken. The satiated trout will take an insect only at longish intervals, allowing many to pass unmolested, making it extremely difficult for the angler to estimate and time the inclinations of the trout, now overtaken by lethargy and exhibiting an attitude surprisingly like that of a ruminating cow.

Very often, during the height of the emergence period, the majority of trout will be discovered in this frame of mind and it is best to abandon them to look for trout on the edges of the hatch, in the weaker lines of drift, where they are getting "skimmed milk"—in other words, fewer insects in the wider currents. They are much more impressionable than the glutted creatures who are no longer able or willing to respond to any enticement.

For the most part the Yellow Breeches does not have the cleared banks and low-lying meadows which are typical of the Letort and Big Springs. The immediate edge of the stream, except in a few stretches, is thickly grown with shrubbery of one form or another, often interspersed with trees, all of which are obstructive to a proper back cast. There is no opportunity here to wander about the close-cropped plushy meadows in low boots,

111

making an occasional stalk and throwing a fly from a crouched position as can be done on the Letort. To reach a feeding fish it is nearly always imperative to don waders (armpit length, for the stream is uniformly deep) and then maneuver about to find clearance behind and before. Even with the assistance of waders, the angler must confine himself to movements near the middle of the stream, more or less, since both banks are usually overhung by the aforementioned shrubbery and the branches of trees, whose tips sweep the surface and oftentimes drag downstream, partly submerged. Such trees create a tunnel-like passage or archway next to the bank, where, if the force of the current is concentrated, it is kept clear and free of any debris and the branches are split and remain parted to permit an entry on the upstream side and an exit on the downstream side.

If this picture is clear, it can be readily understood how an angler, standing in midstream and casting to either bank, must be confronted with the almost impossible task of reaching a feeding trout underneath the branches, i.e., under the archway, by casting across-stream from the center. Though the cast be delivered with the maximum slack in the leader, it will enter the upstream side of the arch and float for only a short distance, beginning to drag and remaining fixed when the drifting leader or line catches in the sweeping tips of the overhanging branches. That can be a serious problem on the Yellow Breeches, aggravated many, many times on those days when the long line of Hendrickson duns are driven against either bank, passing underneath the intermittent archways occupied by feeding trout. If a fish is stationed so far underneath that a short float will not reach him, the first cast that is caught in the overhanging tips will only result in putting him down. A succession of these disappointments is fair cause for avoiding these

112

The Yellow Breeches—looking upstream at the long smooth run that breaks away from the railroad at Brandtsville and flows toward William's Grove. This area, widely known as The Stretch, was probably the finest piece of fly water on the entire thirty-five miles of the Yellow Breeches. Hendrickson, sulphur, and caddis hatches were phenomenal.

The Stretch—looking downstream from the same spot shown in the previous photograph. This portion is perceptibly rougher than that water, and both photographs depict the alternate rough and smooth areas as well as the heavily wooded banks that characterize the Yellow Breeches for its entire length.

fish if others can be found feeding in a clear space or in midstream. Such, in fact, was the attitude that I adopted and practiced for a long time, until one day when I could find no other feeding trout except those that were engaged with the Hendrickson in these difficult places. Feeling somewhat reluctant about abandoning them and disturbed by the prospect of a fishless day, I sat down to ponder the matter and finally conceived what I believed was a feasible plan. A wet-fly fisherman would have had an easy time of it, simply by sinking his fly in midstream then bringing it across and underneath the dragging branches to the trout from a position upstream. With a somewhat similar intent, I moved carefully to a spot some forty or fifty feet above the trout and as close to his bank as possible. I made a cast downstream and toward the center, where I halted it and allowed it to drag on the surface but ten or twelve feet above the trout. Then I dragged the fly across-stream toward the trout's bank until it was exactly poised in front of the entry to the archway, on the line of drift that carried the Hendricksons. I allowed the fly to hang there for a moment, then brought the rod tip down smartly from a perpendicular position to throw a large, loose coil of line in front of me. Immediately upon striking the water the loose coil permitted the fly to be released on its way and begin a dragless float into the archway, where it was taken by a trout.

It was a cast compounded of downstream drift, crossing drag, and a half-roll, so dubbed because a full-roll cast would have lifted the fly off the water. In bringing the rod point down, only enough force should be used to lay a large loose coil of line on the water without moving the fly from its ready position to enter the archway. Immensely pleased with my initial effort, I continued to fish downstream, moving cautiously in order not to send out

113

any alarming ripples, and trying to spot any feeding fish as far away as possible so that I might be able to halt in time to prevent giving away my presence.

I had a singular experience with two trout tenanting the same arch. Both of them were taking well, but that the upper one—that is, the one nearest to me—was the better of the two was plainly evident by the depth of sound that he made upon rising to the naturals, and I was anxious that he should be the one to take the artificial. Proceeding as before, I cast the fly to the center of the stream, dragged it across with the rod held high, adjusted to hang it in front of the arch, made the half-roll, and watched the imitation begin its orderly course downstream. Contrary to my expectation, it was not taken by the first fish but continued onward, where to my surprise it was taken confidently by the second, or lesser fish. Somewhat apprehensive of the possibility of putting down the better fish, I attempted to bring the hooked fish out of his security by first working the line underneath the overhanging branches and then quickly applying pressure to force the trout toward the center of the stream. During this short interval the fly was retained by the trout and the tightened line discovered him soundly hooked and surging toward midstream, free from all obstructions. Within a short time, attended by a minimum of commotion, I brought him upstream, netted him, released him, and returned my attentions to the better fish.

I had not long to wait, for he began to feed again, apparently unaffected by my delicate venture with his messmate. I allowed him to come to several naturals in order to establish his confidence, then repeated the cast described above; the imitation was taken without hesitation. Unhampered by my former restraints, I managed him with considerably more assurance and finally

114

brought him to net, where he was revealed as one of the hold-over trout of admirable proportions, stocked in a previous year but having acquired in the meantime the deep girth and flaming hue of the trout that feed and are nurtured in the rich limestone waters. Him I released too, in accordance with a self-imposed rule; for there is no longer a legitimate surplus of his kind in these waters, the usual catch being something less handsome and more like the product of an artificial hatchery, taken from the water in the same year or maybe the same week that they are planted.

In addition to the directions for cast given above, there is a bit of information connected with the control of the line when the fly has been refused and has passed the trout. Obviously there is a critical limit on the capacity of the fly to float very long on a downstream drift. Before long it stops and begins to drag heavily, putting the fish down and making it useless to try him again. When this had happened to me a number of times, I took care thereafter, before drag began, to ease the leader and fly into the bank as closely as possible by moving the rod tip over and allowing the current to press against the belly thus formed, sweeping line, leader and fly away from the trout with a gentle and subtle movement unnoticeable to the trout. Then the whole assembly may be drawn upstream slowly and carefully along the very edge of the bank until it is clear of the probable orbit of alarm. It is a very pretty maneuver, and repays the angler with opportunities to cast again to the same fish.

These tactics I pursued for two consecutive days, tolling eleven good fish, not many, it is true, but nevertheless the source of highest gratification to me. On the afternoon of the second day I got into the strongest, shrewdest fish that I have ever met up to the present time, though he was not unusually large, not quite two pounds.

115

I had reached the lower end of a long smooth stretch which ended in front of a log lying across the stream from bank to bank. Below the log began a long series of fast riffles, rather shallow and providing uncertain footing on cobblelike stones. I had been standing very quietly in the flat water above the log, looking upstream, when suddenly a fish rose to my left, not more than eight or ten feet away. I had caught the rise out of the corner of my left eye, and without daring to move my head I moved my rod point to the right with stealthy care and allowed the fly to drop from my fingers with the greatest modesty of display; then a sharp little flick of the rod tip from left to right, blocked from the trout's view by my body, carried the looping line from right to left over my head, depositing the fly exactly in front of the rising position. The trout rose, fastened, and immediately tore downstream behind me, passed under the log and into the fast water below, putting a severe strain on my terminal tackle. I turned and followed as quickly as my cumbersome waders would allow, reached the log, passed the rod underneath, and recovered it from the lower side. All the while my reel was proclaiming its discordant objections.

All the strain that I dared to put on the fleeing trout, aided by the current, would not stop him and I was forced to follow, stumbling and sloshing, each footstep a perilous undertaking. At last he stopped, many yards below the starting point, turning to face the current while I hastened downstream wide of his position in order to get below him. It is always best to get below a strong fish and make him work upstream against the current, thereby relieving the strain on tackle and causing the fish to tire more easily. As if he had anticipated my design, he turned and bored downstream again when I had hardly come abreast of him, finally stopping and

116

turning to face me as before. This contretemps was repeated five or six times, taking us farther downstream each time, until I was willing to admit that this was the craftiest exhibition I had ever seen by a trout. When we reached the quiet water below, it was much easier to control him since he then lacked the advantage of the swift current, and bring him to net, where I found that the hook was holding to the merest thread of the tough mouth tissue.

A notable sidelight connected with these events is the fact that I had used only one fly for all of the eleven fish, not a record, of course, but nevertheless pleasing to me because it was strong evidence of this pattern's durability and effectiveness. I have kept it as a souvenir, though it looks no less attractive for so much wear and tear, the all-important wings remaining as shapely and cocky as ever. In the normal course of things I do not enjoy so much service from a single fly, losing so many of them in the trees, in the grass behind me, and in the fish.

The standard tie for the Hendrickson could not have withstood this punishment any better, and if it had, it would not have maintained its appeal on consecutive presentations to the trout without meticulous care in re-drying and reshaping to recover its form. For many years on these waters the standard pattern has been used and decried for its deficiencies; it is the free opinion of many that it has never been a satisfactory specific for the Hendrickson of the Yellow Breeches. It is hard to believe, however, that Ephemerella Subvaria of the limestone country is any different from that of the freestone waters. Above all, it is hard to reconcile the wispy lemon wood-duck wing of the standard artificial with the sharp, clean outline of the dark slate-blue wings of the natural, a criticism that I dare to advance despite the risk of be-

117

ing confronted by someone, some day, with a natural Hendrickson flaunting a speckled lemon wing!

It is conceivable that lemon wood duck might be a fair deal for the light Cahill, but I wonder if Theodore Gordon ever intended that it should be carried to such a contrasting extreme as is represented in the standard Hendrickson. It is conceded too that the blending of hackle might be a legitimate practice to effect certain shadings in the wing of the natural, but what is to be gained by mixing blue dun hackle with lemon wood duck for the Hendrickson? If the hackle is intended to represent wings, it ought to stand alone and unaided, complete in itself; this would be far more logical than the conventional mixture, although a much weaker and less effective representation than cut and shaped hackle-point wings.

There are an astonishing number of fly-fishermen who like to assert positively that trout do not see wings and that they themselves no longer use winged flies (but what is a hackle or buzz pattern, with its many fibres rising above the hook shank, if not a wing?); and in the same breath they insist that they must have "high-riding" dry flies. "On their toes" is the expression often used, and for what purpose if not to elevate a wing above the surface? Moreover, he will probably be rabid in his demands upon his tier for the best of hackle, emphasizing the need for sparkle, translucency, and other qualities that cannot be appreciated by a trout unless they exist above the hook shank, since those fibres underneath the shank are depressed, bringing them in contact with the water, where they are distorted and neutralized by the light pattern—another good reason for denying the validity of imitating legs.

The term "high-riding" is a misnomer and ought to be replaced with the term "high-winged" as being a more

118

accurate descripton of the requirement for the wings of duns; it does not matter whether they float "on their toes" or not, so long as *the effective height of the wing above the surface is the same as that of the natural!* This much I have proved conclusively for myself, in particular with the imitation of the Hendrickson, which I have tied with a tall, cut and shaped wing and extremely short-fibred hackle to support the fly.

EFFECTIVE WING HEIGHTS

I have argued strenuously too for the use of hackle which is the same color as the wing, even though I am convinced of the hopelessness of obtaining accuracy in this respect, despite my belief in its importance. I have tied the Hendrickson with hackle the same color as the legs of the natural (buff color) and deadly flies they were too, but that was only made possible by the use of ultra lightweight, finely tempered hooks, requiring no more than three turns of hackle to float them on our placid waters. Such hooks as these are generally not reliable, and good ones are hard to find; but when one is found, it is a priceless possession, to be fished with extraordinary care lest its loss should occasion our regret. Three turns of short-fibred buff hackle on these hooks does not materially interfere with the form and color of the big slate-blue wing that I use for the artificial, which on the natural is large and out of proportion to the size of its body.

"Bill" Bennett, flytier extraordinary and ardent proponent of the tall-winged, thorax style of fly, tells an amusing story about this pattern. The incident happened in an earlier day when the new pattern was not generally publicized and when I was still anxious to prove it. Bill had been fishing the Yellow Breeches during the Hendrickson hatch when he came upon an angler fishing his favorite stretch of the Breeches, said angler being intensely absorbed in the problem of tempting several good

119

trout that were feeding industriously on the drifting duns. Bennett sat down on the bank to watch the show. Fly after fly, cast after cast was presented to the fish, who disdained everything, including the standard pattern; then Bennett, experimental minded and sympathetic by nature anyway, offered one of the new patterns to the laboring angler, who accepted it with a doubtful manner, regarding the "beast" with evident suspicion, but nevertheless, probably out of sheer politeness, knotted it to his leader.

His first cast resulted in a rise and a soundly hooked fish. Within a short time he succeeded in raising and hooking several others, after which he waded to the bank, sat down, and examined the fly with concentrated effort, muttering unintelligible things to himself while stroking and nursing the parts with a tenderness that one reserves for treasured possessions. His conversion was complete and wholesale.

I have often wondered how much presentation and not pattern determines the outcome of a cast. Others have wondered too and have taken an absolute position one way or another on his question. I have tried many experiments on this score but do not feel free to offer the results as proof of any contention. There are too many imponderables involved which make any conclusion doubtful and render any position untenable, even though the result is favorable to the contender. To conduct such a test properly you must be sure that each different pattern is presented to the fish in the same manner each time, and that means a consistently good delivery by the operator. A single mistake would ruin the experiment. There is not even any way of judging whether a consistently good delivery is being used, for sometimes there may be just the tiniest bit of drag, not noticeable to the fisherman but enough to make a trout hesitate. Then

120

there is the objectionable presence of the leader itself, sunken or floating, light colored or dark, opaque or translucent; and even sunshine or cloud to operate with different effects on each cast.

There are records of experiments to show that any artificial dropped to the waiting trout without the leader would be taken. There are records of experiments, equally leaderless, in which the trout would accept only one of many patterns.

Once when an argument among the Fly-Fishers' Club members waxed hot and heavy on this question, one of the more witty members completely demoralized the meeting with an account of a related experience, not premeditated by him. While standing on a bridge overlooking a favorite pool, he had his fly box open, a large affair with open compartments and containing many flies, and was trying to select from it a likely pattern. While he was occupied in this fashion, a violent gust of wind blew across the open box, lifting half of its contents in the air and dropping them helter skelter on the water. His dismay was overpowering, since the flies represented a costly investment, but imagine his amazement when there came a sudden and tremendous rise of trout to the manna of leaderless artificials. Blessed with a resilient nature, he shrugged off his loss and hastily descended to the stream, prepared to commit a slaughter among the trout, for what was the small matter of a gross of lost flies when there were trout to be caught, dry style. But not a single, solitary rise did he get with any one of his remaining artificials. He concluded his account with a challenge to anyone to say that the trout had not got the best and most tempting patterns in his box; but for the life of him, he could not remember what they were!

Pattern vs. presentation will not be easily resolved, if ever, though that is no reason for not trying to improve

121

the effectiveness of dry flies. A sound fly in the hands of an artist-fisherman is the deadliest of combinations, but even a sound fly will not supply the deficiences of faulty casting, or the knowledge of a given insect's habits on the water or the trout's manner of taking it. There is a liberal education to be acquired in watching a trout's rise from beginning to end, particularly some of the limestone trout, who could put to shame the vaunted keenness of a Yankee trader; their suspicious souls are mirrored plainly at every stage of their studied inspection, mentally fingering the proffered artificial to determine its genuineness. That is to say, the Hendrickson pattern offered here is no panacea, but it is the best that I have seen or used for the Subvaria hatch.

SPINNER

It is the oddest thing that in all American angling literature so little has been done to establish a better understanding of fishing to spinners. A beginner would find it almost impossible to locate a satisfactory presentation of precept and example for his guidance. Much has been done on this subject by British anglers, but they have left much unsaid about this highly important phase of the dry fly. If the truth were known, and this I submit is the truth, fly-fishing to some insects is almost entirely confined to the evening fall of their spinners, the duns being more or less ignored as a general rule, since there is more concentrated fishing in a fall of spinners than in any hatch of duns.

The reason for this is easy to understand when it is noted that the duns of so many species hatch out at long-spaced intervals during the daytime, so that there never seem to be enough insects on the water at any one time to cause a rise of trout. It is a deceptive thing and may

122

easily create the impression that the hatch is poor; yet when evening comes the fall of spinners of that same species is so great as to resemble a blizzard. All of my experiences with the Green Drake have convinced me that this spectacular insect should be regarded in this light, thereby accounting for the generally poor rise to the duns but the magnificent fishing to the spinners.

In some other genera or species the duns come down in regularly spaced flushes or fleets; and that is something that a book on entomology would not tell the angler. It means that there are many more individuals in the same place at any given time. This situation is more attractive to that severe economist, the trout, who needs to get his food in worth-while quantities; yet, actually, there is not a better hatch of duns than in the case of the Green Drake. In addition to this, there is the habit of some duns of riding the water for a great distance, thereby overlapping the emergence moments of other individuals, so that their numbers on the water at any given time are plural rather than singular.

These are the main reasons—the flush emergence and the lengthy rides—for my not having confounded the reader with a multitude of insects which do not exhibit these desirable characteristics. Ephemerella Subvaria, Ephemerella Invaria, in fact all of the genus Ephemerella which abound in the limestone waters, are outstanding in meeting the requirements of this prescription, and my esteem for the Hendrickson is even greater because he emerges during a season when rough, rainy, cold weather prevails, causing him to remain on the water for a longer period than is ordinary. It explains why a very sparse hatch of the Hendricksons will sometimes provide wonderful dry-fly fishing.

If our considerations were limited to the Hendrickson dun alone, my extravagant praise of him would be amply

123

justified, yet the Hendrickson is equally praiseworthy because of its habits as a spinner, for which it excels, to my knowledge, all other species. They come down at an earlier hour than any other species on the limestone waters, and that is no small favor for the evening fisherman, who is oftentimes desperately pressed for time and daylight when the trout are rising like fury and he cannot see to thread a leader point to his fly.

Midseason and late-season spinners are not nearly so co-operative, being more or less nocturnal in their habits and allowing of comparatively little daylight fishing. Nothing is more galling to the evening fisherman, who, champing and impatient, stands and watches with growing anxiety the great clouds of spinners high in the tree tops, engaged in their seemingly endless dance and dropping to the water ever so slowly while he anxiously computes the remaining minutes of daylight.

The spinner of Ephemerella Subvaria does not offend in this respect but makes its appearance at a rather early hour, as early as five or six o'clock, moving along in precipitate flight, its yellow egg sac cradled underneath its body and its wings beating furiously with an intensity of purpose that is, relatively speaking, as great as that of a charging lion. There is no performance by any mayfly more profoundly moving than the evident singleness of purpose and the undeviating course which this elegant lady pursues in fulfillment of the design decreed by nature. Once having delivered her precious burden into the keeping of the living stream, she falls to the water exhausted and gradually extends her parts until they are lying inert on the surface, thus becoming an object of interest to the trout, who are bent on a more prosaic mission.

Fishing to the spent imago is not confined to that evening when the flight takes place. It is sometimes profitable to use the spent artificial on the following morning, for the spent forms may remain along the edges of the stream, where they have been caught or trapped by twigs, leaves, or the bank itself. Not all of them remain there; some of them are dislodged by a vagary in the current or other influences, causing them to pay out into the lines of drift until they effect another lodgement or are taken by the trout.

The practice of taking water temperatures is a good one and should be related to the morning fishing to spinners. When the waters are sufficiently warmed to cause the trout to begin the search for food, the first rises will probably be to the spent imagos lying in the backwaters or eddies. For a short time, at least until the duns appear, the feeding will continue in this fashion—sporadic, it is true, but nevertheless containing the possibility of some diversion for the angler. Sometimes the fishing to spinners on the day following the flight is more extensive than many realize. It may extend into midday, a phenomenon that is peculiar only to a stream of extremely slow, smooth current or one which is not subject to rapid rise or fall. I am satisfied that the spinners do not tarry long on the northern limestone or freestone waters, where the rapid current and stretches of rough water would quickly dispose of the spent forms long before sunrise; but if the stream has a constant level and a slow flow, the spinners can often be seen on the day following the evening flight.

Three years ago, when the Hendrickson hatch on the Breeches was the finest I have ever seen, the water had reached an unprecedented low level, which was maintained without the slightest variation throughout the hatch, no rain having fallen for the entire period. Normally the Breeches is a very slow-flowing stream anyway, being slowed by a long series of ancient mill dams

125

for almost its entire length, so that its lagging spinners remain along the edge of the stream and in the backwaters for an unusually long time.

It was near the end of the emergence period and near the end of some of the finest dry-fly fishing I have ever enjoyed, but I went to the stream anyway, not hopeful, it is true, for I was sure that the duns had ceased to emerge. Nevertheless I prowled along the bank for some time, and it was somewhere about midafternoon—perhaps three o'clock—when I saw the first rise, very quiet and daintylike, along the edge of the left bank looking upstream, and underneath the spreading branches of a beech tree. The rise was repeated a few more times at irregular intervals, and I was at loss to account for it since there were no duns in sight and only a few caddis, which are not ordinarily taken without being accompanied by the slashing rise form. After puzzling about it for some time, I moved closer to the edge of the stream to examine the water and much to my surprise discovered the spent imagos, lying in the surface film and hitching along from lodgement to lodgement, close to the edge of the bank.

I concluded that the spent artificial was worth a try and got into the water below the beech tree to throw one at the rise that I had seen. It was a bad position, the water being very deep behind me, hence unwadable, and the bank being heavily overgrown, therefore giving no chance to run with a hooked fish; but I threw to the fish anyway, measuring him short with the first and reaching him accurately with the second cast. The fish rose and fastened so quietly that I was not immediately aware that I had got into one of the best trout of the Yellow Breeches until he made a powerful run and continued to burn the reel, making a straight course upstream. I allowed him to have his way, as is my custom

126

after setting the barb, and when he had spent himself somewhat I tightened the line to make him spend himself some more. Evidently he became alarmed, for he made a quick turn and headed downstream, directly toward me, a maneuver of which I had ample warning from the violent hissing of the line. There was nothing to do but kick the water and make a commotion in order to scare him back upstream, but this dodge succeeded only temporarily. He seemed to reflect for a moment, then made another quick decision and headed straight downstream a second time. He was a most determined fish, and I had a suspicion that he would win past me into the impassable water below, but I wanted to see his size; when he went by, I tried to lift him to the surface with all the strain that I dared to put on the tackle but I do not believe I lifted him a single inch from the low level that he was traveling. He reached the deep water below, where I could not follow, and the hook finally came away from the force of trying to hold him. *Sic transit gloria!*

Nothing daunted and fully aroused to the possibility of more rises of this nature, I sat upon the bank under that same beech tree to watch the water, with one eye on the well-worn streamside path, ready to forestall any invader with a pretense of fishing (really, one has to do it in these times), and ready to try my luck again with the spinner.

That was an interesting piece of water underneath and in front of the beech tree, and it finally dawned on me why it should be ideal for rises to overtime spinners. The stream made a bend at this spot and formed a kind of eddy which received all of the drift from a long straight stretch upstream. After a short waiting period I thought I saw a small movement on the surface about 15 feet above the beech tree and very close to my bank, then another, similar movement, 8 or 10 feet upstream but farther out

127

from the bank. There seemed to be two different fish, but I could not be certain. It is sometimes very hard to detect the rise to the sodden imago because of the trivial rise-form. In any event I got into the water underneath the beech, stooping low to avoid the branches, and delivered the fly to the nearest fish, with a side-arm cast. It came off very well, the leader having been allowed to fall before it straightened in order to keep the line on the bank and effect a right curve of the leader to present the fly in front of the trout. There came a quiet rise, a tightening of the line, then a short bit of activity after which the hook came away from a poor hold. He was a fair trout, perhaps 1½ pounds, but not nearly so good as the former.

Still hopeful for another chance, I remained in the water underneath the beech as motionless as possible, while watching the water further upstream for some sign of the other fish whose presence I had suspected. After a considerable wait in this cramped position with no rise forthcoming, I decided to try a few casts on the off chance that he might be induced to rise. I extended the line upstream some 10 or 15 feet and began to search across the eddy, placing my casts carefully in order not to commit an error. When five or six casts had been made, he finally rose, was hooked, and was duly landed, a fish of probably one pound, which I returned to the water. There appeared to be no more prospects that afternoon so I quit the stream, entirely gratified with the outcome of what had seemed like a fishless day in the beginning. My luck had been poor with the two better fish; yet not poor, for in the lexicon of the fly-fishermen, the words "rise" and "hooked" connote the successful and desirable climax; landing a fish is purely anticlimax.

Since that time I have given that particular eddy my

special attention in each succeeding season, with the result that I have experienced similar satisfactory spinner fishing. It has become one of my favorite haunts, and I am sure that in it lives a tremendous trout with whom, someday, I shall cross barbs again.

Fishing to spinners, like fishing to duns, needs to be practiced with special regard for the habits of the insects on the water and the habits of the trout in taking them. That is the basic formula for fishing to any insect. I have elsewhere stated that limestone trout, as a rule, do not ordinarily hover close to the surface when taking duns; this sometimes happens at the Paradise on Spring Creek, but that is the only exception that I can recall. On the other hand, spinners are taken from a position very close to the surface when the evening fall takes place. I am not sure whether a trout comes out in the evening before the fall of spinners or whether the fall brings him out to feed. In either event he is ready and waiting to get the most food in the shortest space of time that he can manage. It is an attitude and a position exactly suited to engage the spinners, whose sudden and simultaneous presence on the surface supplies food in great mass but food that is quickly washed away, so that a trout must economize his movements in order to obtain a fair share of the individual spinners. Here again are demonstrated the correct adjustments which a trout makes in order to satisfy his biological needs, and, as before, he regulates his own pace to that of the insect.

The descent of the spinners en masse does not permit the trout to employ the languid drift and rise with which he intercepts that gentlemanly fellow, the Hendrickson dun. With the spinners he must become very busy, lest when the evening's account is cast it shall find him poorly gained, as it will if he has been sluggish. It is this same

129

busyness of his that creates so much difficulty for the fly-fisherman, who sometimes becomes confused, terribly excited, and completely disorganized by the churning activity of the feeding trout. Despite appearances it is really an orderly affair, as it should be, for order manifested by well-marked and logical sequences is the law of nature, a law no less applicable to the mad scramble for spinners than to anything else, though one would not believe it from the effect that it creates on the flailing angler.

Observe carefully the habits, not of many trout, but of one trout alone during this melee. He is hanging close to the surface. He makes a short stroke to the right, then one to the left, another straight ahead, then back to the right and left again acquiring in this fashion five or six spinners in very quick order. There follows a blank interval when he drops down and remains motionless for a few moments while he masticates or swallows, or whatever it is that a trout does to get the insects into his gullet. Then he rises again and makes another quick killing with the same procedure, repeating it as often as he can until the spinners are washed away and he can find no more.

There are a number of strong inferences to be made here. First, it is likely that he is almost entirely independent of his "window" for the purpose of seeing his food, since he is so close to the surface that the window's area is considerably reduced. This conclusion is supported by the fact that he usually moves forward, indicating that he is being lured by the light pattern or, more correctly, the body of the spinner on the underside of the mirror beyond the circumference of the window. Next, it is clear that lines of drift are not rigidly observed, because he is shifting from side to side to gather

130

the pressing spinners. Lastly, it is plain that long floats are out of order because he is not waiting for a float but goes forward a little to meet the insect; this is fortunate for the angler, since long floats are time consuming and might cover only barren water for 90 per cent of their length.

The best possible procedure, then, to meet these conditions is for the angler to move as close to the trout as possible without alarming him and by using a short line make numerous casts across the line of drift in front of the trout's position, bracketing this area with short floats (but not too short or too often, else the trout does not have time to take the fly). In the earliest stages of the spinner fall these pecularities are not so pronounced, nor when the fall is very sparse, as one would expect when comparatively few insects are on the water. This is especially true of the lagging spinners which are sometimes taken on the following morning but taken at long spaced intervals in the backwaters or slow eddies therefore requiring a more deliberate style of casting in a stricter line of drift or to the spot where the trout was seen to make his last rise.

BRACKETING

In the chapter on fly dressing some novel suggestions are offered concerning the construction of the spinner to conform to the habits of the different species. Now I will make another pertinent observation, one that may seem exaggerated at first blush, but is nevertheless confidently given as the startling truth, and it is this: that no matter what the ultimate position of the exhausted spinner may be, in the early stage and during the height of the fall, a trout rarely or never gets a spinner in the flat-winged position that is traditionally prescribed for the imago. This I have confirmed again and again by creeping close to a trout near the bank and seeing him busily take the spin-

131

ners striking the water in front of him, with wings upright but not closed at the tips. What is more, they do not touch the water as daintily as their frail bodies would lead one to believe. There is a definite little splash which must certainly demand the trout's attention and intensify the light pattern; consequently the artificial ought to be splashed a little.

For these reasons I have adopted and recommend the plan of tying the spinners in two styles, one a half-spent pattern with wings partially elevated and the other full-spent to imitate the spinner in the last stages when it is fully prone, both wings flush with the surface. There is a great difference in the light patterns that these two types create.

In the case of the Hendrickson spinner only the half-spent pattern need be tied, since this is one of the mayfly spinners that falls on its side, one wing on the surface and one in the air. This position is very easy to achieve with the artificial, simply by tying the wings fairly upright but with no hackle, or two turns at most, for support. It will fall over on its side with almost every cast, and it is the most killing fly that I have ever used for the Hendrickson spinner.

There was a time when I believed that it was necessary to tie the wings of spinners with the iridescent colors displayed by the natural—wine red, green, blue, and bronze —something which can be done by mixing hackle fibres in these colors; there is good reason, however, to suspect that the trout never see this iridescence. This play of colors can be seen by any observer who is looking down on the spinner from above because it is reflected back to him, but the colors are not transmitted through the wings to the trout who is seeing the spinner from below. If there were a strong light striking upward from the trout's

132

position, the flutings on the underside of the spinner's wings which cause the iridescence might also be reflected back to him. That is the way it appears to me after careful study of the matter.

Chapter 5

BLUE- AND PALE-WINGED SULPHURS

So GREAT is the fascination of the Hendrickson for the fly-fisherman, so complete is its well-rounded appeal of day-hatching dun and evening-falling spinner, that when the last of the big blue wings has fluttered away into the treetops, when the last of the spinners has been washed away into oblivion or disappeared down the sucking throat of a trout, there comes a feeling as though the end of the season had arrived, a feeling of finality as though the great highlight of a fly-fishing year had been reached and passed, as though it were time to put away the rod. Lot's wife did not cast a more lingering backward glance in the direction of nostalgic memories than the Yellow Breeches fisherman who has lived a full ex-

135

perience with this hatch at its best; and yet the season has hardly begun. It is hardly the beginning of May, when the erratic awakening of spring elsewhere has stirred the fly-fisherman to assemble his gear and begin his adventures anew with the trout.

True enough there are other early-season hatches and waters that afford some fly-fishing, but these are nothing compared to the Yellow Breeches and its Hendrickson. Boiling Springs Lake, a small artificial limestone pond, sometimes surprises the early season angler with a fine show of a large mottled mayfly, Callibaetis Pretiosus Banks, which favors weedy, still waters like this and the millpond at the head of Big Spring at Newville. Neither of these places, even though they are dependable for the Callibaetis dun and rising fish, can be regarded as anything more than a second choice when the Yellow Breeches is the color of its name or a late spring keeps its water too cool for the trout to become active. Both of these ponds are too small and are often too crowded with anglers to be attractive to the serious fly-fisherman except in late season, when they are abandoned by all save a small group of first-rate fly-fishermen from the vicinity of Carlisle, who are especially faithful to the charms of Boiling Springs and understand its moods better than anyone else.

In any event, with or without a good Hendrickson season behind him, the Yellow Breeches angler must suffer a period of blankness, really a very short period, not more than a week or two, but seeming so much longer for one who has been using the dry fly with fervor as though it were Green Drake time in June. How pleasant then to hear, perhaps at the end of the first week in May, that the blue-winged sulphur is beginning to emerge. The angler prepares to sally forth again to continue the

136

pursuit of the hatches, no less enthusiastic than when he anticipated the Hendrickson.

But it is difficult to speak or write of the sulphurs without associating them with Cedar Run, synonymous terms for those who have trod the banks and learned the secrets of this, a little gem of a limestone stream whose brilliant facets glitter with the light of its many powerful but obscure attractions. If you happened to meet such a stream as this on a casual journey, it would scarcely merit more than ordinary attention, for its greatest virtues are more or less hidden, cloaked by its miniature aspect and a rather barren appearance (it is not nearly so fertile-looking as the Letort, Big Springs, or the Yellow Breeches—good fat water not excelled anywhere). It is a mistake that no one should ever commit on first acquaintance with any limestone stream, for such a stream is never to be idly dismissed, no matter how small or insignificant it seems—and least of all if it is Cedar Run.

If anything is really constant in this changing world, constant as the stars that endure, as the sun that shines, or the seasons that turn, it is the blue-winged sulphur of Cedar Run, a beautiful mayfly, regular in its annual appearance, sustained in its daily presence for five or six weeks, its existence inextricably bound to the romantic history of this little stream.

Many years ago when the brook trout held sway, Cedar Run was discovered by a breed of fly-fishermen, now extinct, whose names have become legendary but who fished wet style entirely, a burlesque thing in this stream where trout rise so freely to the blue-winged sulphur and its spinners. Somehow, though, they made great catches with their wet flies, worms, and minnows. Twenty or thirty brook trout to a single fisherman in one afternoon was quite common in the old days; many such catches attained an average length of sixteen inches or

137

better and an average weight of two pounds or more; all of this in a little limestone stream not wider than twenty feet and with an average depth of two or three feet, often less in times of drouth.*

What a picture they must have made, those early-day fishermen in their frock coats, high hats, and starched collars, one hand plying a "weepy" rod and the other hand clutching a huge, long-handled net as a support for their stalwart mid-Victorian rigidity, an attitude to be tolerated only by the naïve brook trout of olden times. One of them I saw some years ago, the last of his kind, moving slowly along the smooth bank of the Letort, each dignified pace marked with a vigorous thud of the net handle and a simultaneous flick into the calm waters with his team of wet flies. It must have been for their benefit that Charles Kingsley of *Water Babies* fame wrote his caustic injunction to "take off that absurd black chimney-pot . . . crawl up on three legs and when you are in position, kneel down. So." What a marked contrast there is with the modern angler, now compactly accoutered, efficiently tooled, and more circumspect in his approach; but what a marked contrast in the numbers of fish that he gets, nothing like the great baskets of trout that grandsire secured. Of course, grandsire did not have to deal with that newcomer, the brown trout, whose skulkings and shyings

---

* It is not so long ago that the dry fly has been introduced on these waters, hardly more than a quarter of a century. Gleanings of an angler-researcher from some of the old downstream wet-fly school will oftentimes reveal a curious attitude where the hatch is concerned. Once, when one of them was questioned about the best method of fishing to surface rises in the old days, he replied that when the hatch and the rise were in progress it was customary to find a convenient place to sit and watch until the activity abated, then to resume fishing with the wet fly—for everyone knew that trout could not be taken when they were feeding from the surface. What robust characters these men had! The hatch was no problem at all—they simply ignored it!

138

and coyness are trebled by the daily tramp of countless footsteps near his lair (and nowhere has he effected a greater change than in Cedar Run).

As nearly as can be determined, brown trout were first introduced into Cedar Run about the year 1930 in a most romantic way. It seems that a Mr. O'Malley, then head of the Federal Fish Hatcheries, accompanied by another person, came to Harrisburg to visit a Mr. Gus Steinmetz and all together they made a survey of Cedar Run. It was Mr. O'Malley's opinion that Cedar Run was ideally suited for the propagation of Loch Leven trout. A short time after the visit Mr. Steinmetz received a call from Washington from a caller who identified himself as President Hoover. Mr. Steinmetz, doubtful and suspecting a prank, could hardly restrain an impulse to make a ribald answer; fortunately he did refrain, because it actually was the President, who wanted to come to Pennsylvania and go fishing on the following week end. For some reason, the trip never came off, but shortly thereafter a call was received by Mr. Steinmetz that twelve cans containing Loch Leven trout would arrive on a certain train, at a certain time. The cans arrived in due time, bearing a total of some 150 trout six inches or longer which were forthwith placed in Cedar Run. If the good President had anything to do with this shipment, it would have pleased him to know that it marked the beginning of a new era for Cedar Run, a new lease on life, as it were, to take the place of the ancient brook trout, who could not survive the hardships of changing times, to say nothing of his greatest weakness, his forthright and gullible nature.

From this humble beginning the brown of Loch Leven (I believe they are the same species) of this stream multiplied with unbelievable rapidity and grew to an enormous size the like of which one would never expect

139

to see in a stream so small as this. Four and five pounds is not unusual; though I have never yet taken one so large I have seen them caught, and one that was recently taken was close to eight pounds. Such a thing is possible only where there is an adequate food supply, and Cedar Run surpasses anything of its kind in this respect, as anyone can see, even with a cursory examination of the stream bottom, always creeping and crawling with shrimp, larva, and crustaceans of every order, abounding in astronomical numbers. It represents a capacity not only to grow trout to a large size but also to grow them in quantity.

For many years the upper half of Cedar Run has been closed to the public by riparian owners; consequently comparatively little fishing is done today, with the result that this section has been a wonderful spawning area and a safe refuge for the nursery stock, and has faithfully supplied the lower reaches with a yearly crop of fingerlings and larger fish who have outgrown their meager surroundings, though really if there is enough food a trout does not need much water.

Best of all, this particular breed of Salmo Fario is a free riser and this, joined with the allure of the abundant blue-winged sulphur, explains the eager and attentive presence of the dry-fly fisherman from the first week in May until the second week in June, excepting those interludes when the hue and cry over the Green Drake has penetrated his deep absorption and he becomes torn with opposing desires. It is likely that he will make a flying trip to the northern limestone streams, Spring Creek, Spruce Creek, Fishing Creek, and the others, and that there he will renew his conviction that the Green Drake is a hoax—an overrated, puffed-up monstrosity that isn't worth even a few minutes of the precious time that he has lost with the blue-winged sulphurs. He is

140

likely also to discover anew that the blue-winged sulphur of the northern waters, emerging at the same time as the Green Drake, is just as fine, as interesting, and as attractive as his own Ephemerella of the southern streams. I have enjoyed some rare times in this unexpected fashion, times when Ephemera Guttulata was a sore disappointment and I was glad to spend my precious weekend fishing to the sulphurs as I would have done at home.

In fact, the fly-fisherman can visit any of the limestone waters from the meanest to the largest during the latter half of May and the early part of June without risk of hatchless days, for this Ephemerella and several lesser ones thrive in every one of them. They are the great staff, the pillar upon which every limestone fisherman leans for the bulk of his dry-fly fishing and believe it, for the southern fisherman it takes a mayfly of this sort to dispel the mesmerism of the earlier Hendrickson.

Theodore Gordon knew them well. The great man has recorded some dramatic and baffling moments with them at Spring Creek, Bellefonte, Pennsylvania. In years gone by there used to be a wonderful hatch of the sulphur in late summer on the Letort, but for some reason it has sadly depreciated though other Ephemerids are still in abundance. It can be found at Newville and on the Yellow Breeches too, but on the latter it is an unpredictable affair, alternately good and bad in different seasons (when it is good, there is no better hatch anywhere). May 9 of the year 1945 was the occasion of the most stupendous hatch of the sulphurs ever witnessed by local anglers on the Yellow Breeches. It is likely that those who visited this stream on that bright cool day will never forget the awe-inspiring sight that they beheld. A strong wind was blowing downstream and great clouds of these mayflies were sent billowing before the breeze. Intermittent gusts of wind started new clouds, one rolling

141

upon another, intermingling them in a bewildering mass of glinting, golden bodies and light-struck wings. Sad to relate, the fishing was extremely poor that day, the head of trout in this stream having deteriorated so badly during that period that only six small trout could be seen surface feeding. Four of them were caught and returned, the rest of the day being spent in observing the unusual and spectacular appearance of this hatch, an occupation of small comfort to the fly-fisherman, wandering along the banks in a disconsolate state of mind.

On Cedar Run it is different. There are no such temperamental outbursts in one year and scarcity in the next. It has always been consistently good, for Cedar Run, little marvel of a stream, harbors and nurtures the sulphurs in generous numbers, pouring them out unfailingly, season after season, as day-hatching duns and receiving them evening after evening, as spinners, their ranks seemingly unbroken by the great toll taken by the trout and the birds, those gourmands who never seem to get enough of them. How many hazards this dainty mayfly must escape to fulfill its meager destiny; to live the few fleeting hours of sunshine and air, of love and life. It is bad enough to risk immediate destruction by a trout or the dangers of wind and storm or frosts and flood, but worst of all, there are the birds, ganged and perched in fierce array, watching and waiting with the keenest eyes of the animal kingdom to guide them unerringly in a swooping flight that surely spells doom for the laboring dun, freshly emerged, beating his way upward. As though some telegraphic connection advised them, as though some beacon directed them, they wheel into the streamside trees and foliage in singles and doubles, in squadrons and battalions, twittering and chirping their foul greed to one another in raucous tones that dispel the glamour with which they are too often invested.

142

Let a single dun survive the hunger of a trout, the treachery of the currents; let him dare to begin his upward flight, and it is barely started before there comes a flutter of wings, a flashing dive, and he meets disaster. Even though his pursuer should miscalculate and the dun escape for a moment, there comes another who, equally swift, has been poised and waiting, ready to rob the first would-be executioner; and if the second misses in turn, there is yet another, and another, to seal his doom. But they do not often miss, since if the first has not committed himself too far in his dive he can turn swiftly for another stroke, turning again if need be, meeting each dodge of the mayfly with a complimentary movement, pressing nearer and nearer until he can close with his beak. So it goes hour after hour, mayfly after mayfly, in a grisly feast that leaves the birds panting and choking, surfeited beyond all reason, yet they must swoop for another.

Has anyone ever autopsied a bird, a small bird, to see how many mayflies he can eat? I know by actual count that a one-pound trout can eat one hundred and sixty-eight sulphurs at a single sitting—one hundred and sixty-nine if the artificial which lured him be counted—and he would probably have taken more! It is a wonder that these fragile Ephemerids have not disappeared from our waters long ago; if they have not, it is only because there are so many of them, for they have no other defense.

On some of the limestone waters there are outcroppings or ledges in the stream bed that create a sharp drop and a short stretch of swift, white-capped water like that which is often found on freestone streams. They are a deathtrap for thousands upon thousands of these frail creatures. I have stood at the lower end of these stretches and watched the sulphurs bursting out of the white water to glide onto the smooth surface below, wings

143

sodden and body limp, never to rise again and never to author the progeny of their kind. It is a matter of good husbandry, wherever control can be exercised, to alter the character of these parts, to turn them into smooth glides or slicks and thereby obviate any further embroilment of wings and bodies with the rough water. How many of these insects must be destroyed each year under these circumstances, yet their destruction goes unnoticed, although it undoubtedly contributes to the ruin of the hatches. It is doubly worse in a period of bad weather, when the duns remain on the water longer than usual and do not rise successfully before they are swamped by the white caps. Such needless loss as this could be easily avoided where proper management is feasible. I have often promised myself that, if given the opportunity, I would devise some scheme and put it into operation to correct a condition of this nature.

Such an opportunity came a short time ago when one of the most desirable parts of Cedar Run was purchased by Mr. Henry Clayton of the Harrisburg Fly-Fishers, and I was immensely pleased when he invited his friends to make suggestions for the improvement of this piece. I had long deplored the existence of several of these traps on this stretch and therefore urged the construction of alternate baffles to slow the fast water and throw it from side to side in a kind of switchback, smoothing the water considerably and allowing the duns to negotiate the trap unharmed. The baffles that were formed benefited the stream in many other surprising ways. Instead of constructing them as a straight arm out from the banks and slanting downstream, a hook was added at the end of the arm, curling upstream to catch the flow and raise the height of the water, which has always been shallow in Cedar Run. These newly-named hook dams (for they are really shaped like a fishhook, with the eye anchored at

144

the bank) served an additional purpose since they functioned as a resting and hiding place for the trout on the concave inner side, and this was further insured by placing a cover of logs and boards over the top. But far and above any other benefit, this unusual construction multiplied many times the opportunities for the best kind of dry-fly fishing, something that the reader would never understand unless a more detailed account of dry-fly fishing on this stream is given. Such an account as I shall now give will help to clarify the importance of the hook dams on this stream, at least, and may possibly indicate how they could be used on other streams too.

I have elsewhere intimated that Cedar Run is an obscure piece of water, to be studied and learned, courted and considered, if it is to be fished well. One of its strange ways is the peculiar manner in which the trout feed. This is something which is entirely governed by the physical aspect of the stream itself. For the most part, in truth, the best part, it is narrow, shallow, and less weedy than others, and it has a steeper fall than the ordinary limestone stream. Its banks are heavily wooded, another unusual feature, and periodically, because of wind and storm, many branches, dead trees, and other forms of debris fall into the water, float for a short distance, and are trapped to form a miniature log jam,  which yields to the force of the current to form a concave side exactly as it was duplicated in the hook dams. This concave side is nearly always deeply undercut, and here is where the trout can be found, large and small, each succeeding jam containing a like quota, so that almost the entire trout population is disposed underneath these shelters.

These jams do not always fill the stream from bank to bank, the majority of them having a narrow break or gut, forced open and kept clear, close to either bank. Now it

145

is true that the trout do not always stay underneath the jams, for some of the smaller ones often come out to feed in the shallow clear runs above them, but the larger fish, as a rule, remain under the jam and take a floating dun from that concealment, thereby creating a most unusual problem for the fly-fisherman.

It is the most fantastic thing in the world to see a cocky blue-winged sulphur float downstream, drift into the face of the jam, bank slowly with the current and travel along the edge until it reaches the deepest bend of the concave face. There it stops and eddies about for the longest time, turning around and around, the circumference of each revolution becoming a little larger each time until finally it is seized by the outlying positive currents and carried around the end of the jam and through the narrow gut to the stream below. In this way, thousands of duns are paraded in front of the trout, where they obligingly stop, pirouette, and are sampled by a breed of fish who might have read Horace and learned about Roman orgies, for they seem to lean on their elbows and allow the food to fall into their mouths. No matter how often the angler sees it, the rise to the slowly eddying duns is always a vast surprise. The trout cannot be seen or heard; there is not even a rise form to indicate the rise. The dun simply vanishes, and there is only the faintest hint of a little half-moon ripple glancing away from the face of the jam.

I have never clocked the time of a dun's stop in front of a jam, but I am reasonably certain that sometimes it lasts as long as five minutes before it is taken. Only once before have I seen anything to compare with this freakish behavior and I think it is worth telling. It happened long ago in those halcyon days when I prowled the foothills of the Alleghenies in western Pennsylvania, fishing the little mountain streams with the dry fly, which was very

146

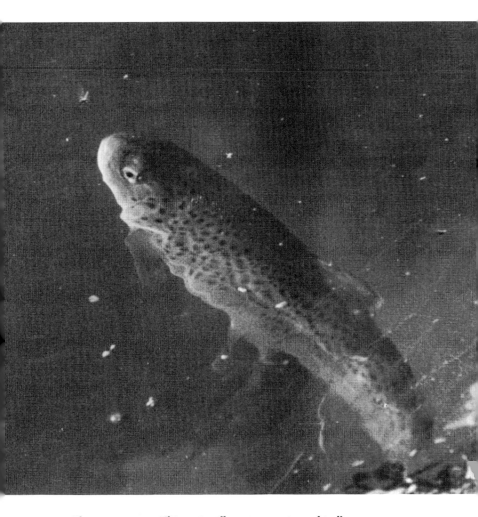

The sipping rise. This series illustrates most graphically the gentle movement and minimal disturbance created by a trout in this type of rise. This trout had his observation post in a dark eddying corner of the Letort, under a pile of debris in the lower right-hand corner of the photograph. He rose all day long, unnoticed and unmolested by a parade of fishermen. He is rising to take a tiny mayfly about to pass over him.

A fantastically lucky and exquisitely timed photograph of the exact moment when the insect is passing over the trout's lower jaw and falling into his mouth.

The turn downward immediately after the insect has been taken. This is a gentle, porpoising kind of movement that causes the surface of the water to bulge and wrinkle, creating a riseform very similar to the disturbance and movement of a nymphing trout.

The end of the rise. The trout is slowly settling back to his original position under the debris, while the tiny wrinkles and bulges drift slowly away over his back.

new to me at that time. I stumbled upon a pretty little stream embedded in a deep gorge the name of which I cannot reveal for sentimental reasons. It was one of those little burns with alternate thin trickles and deep, round basins with no perceptible current in them, mirror smooth and clear as crystal—none clearer. I saw some very quiet rises on one of these basins and tried very hard to interest these fish with my dry fly. Cast after cast was made to these fish, with an appreciable wait of five or ten seconds between casts, but no rise was forthcoming. Thinking that perhaps my tackle was not correct, I refined my leader, changed my fly, and lay to with long casts in order not to alarm the trout. Still no results. Then I had an inspiration; I called up my fishing companion  of those days and asked him to go up to the pool, crawl to the edge of the bank, using the bushes for screen, and see if he could discover anything important about the behavior of the trout in that pool. He performed this maneuver very well, getting to the edge of the bank Indian fashion without alarming the trout. I began to cast again, allowing the fly to drift for a few seconds and was about to make the pick-up when he said, "Hold it." I held, then decided to lift, but he said "Hold it" again. A full minute must have elapsed during this time, and I thought that my friend was crazy and told him so. Finally I could stand it no longer, feeling very foolish about the whole business, and asked him for an explanation. He replied that a trout had been in the act of rising ever since I had first made the cast. I couldn't believe it, but within a moment or two the rise occurred to my fly and my friend was grinning and chortling in high glee.

We tried it again and again all that afternoon, getting a heavy basket that way; I got to the point where I simply made a cast, put the rod in the crook of my arm, and waited until the fly was taken. I asked my friend for

147

full details later on, since he saw more than I, and he explained that the trout just ambled along on the bottom until they were under the fly, then floated upward so slowly that he couldn't see any movement of fins or tail.

The sum total of these and many similar incidents makes a strong case for the proposition that trout adjust their feeding pace to the habits of the insect and the water in which they live. I have always found it to be a sound principle to follow. Your trout is a metronomic creature: rhythm is his creed and he will insist on using his own special beat. It is rarely the same for two streams, even with the same species of trout taking the same species of insect and sometimes it varies with individual trout in the same stream, not more than a few yards apart.

The sulphurs are notoriously fragile insects; they are not able to pass over water that is the least bit rough—water that a Green Drake could ride without ruffling a single one of his setae. If a trout is feeding in one of these stretches that have only mild ripples where the sulphurs are hatching, he seems to hurry a little, making quick sharp rises to take the sulphurs, who are thrashing about on the surface, trying to emerge from the nymphal shuck and having a great deal of trouble in the process. Even after emerging they are not quite so efficient as the Hendrickson in rising from the water. Sometimes they will essay one or more attempts to rise from the surface only to fall back in a welter of thrashing wings, bodies, and legs, then right themselves and try again. On some of these rough runs the trout can be seen taking the struggling sulphurs time after time, usually before the insects' wings are fully drawn from their cases and elevated. A wingless pattern, sparsely hackled, splashed in front of the trout, would take them almost every time. Yet in the quiet waters where the dun is nicely composed

148

with wings aloft, it needs a pattern with convincing wings, hacklewise or otherwise, and it must be floated for a longer time to accord with the deliberation of the trout.

Queer things can happen with a properly winged dry fly. One of my friends, who ties the sulphur with a sky-blue, cut and shaped hacklepoint wing, likes to recount his first experience with this pattern. He found a fine trout feeding in a narrow drift on the sulphurs of Cedar Run. One of the new patterns was knotted to his leader point and thrown to the trout, but the cast flew wide of the mark, some two feet or so to the left of the trout. Nevertheless it was a good cast that floated dragless and serene with wings nicely cocked, but he despaired of it and was about to lift when to his astonishment the trout abandoned his line of drift and slid over to the left to meet and suck the artificial. That is a most extraordinary thing for a trout to do when he is concentrating on an established line of drift, though he will often do it at the beginning of a hatch.

The problems that were posed by trout feeding under the log jams were concerned mostly with the task of a proper presentation of the fly. The prime objective in every instance is to get the artificial to stop and eddy in front of the jam. It is almost impossible to do this by casting across-stream or quartering down, since the intervening currents quickly seize the leader and line, causing the fly to be dragged away from the jam before it can remain long enough for a trout to take it. Not even the overland cast, prescribed in Chapter 2 for the slow eddies, will accomplish the desired purpose. The best method that I know is to go behind the jam, if possible, on the downstream side and cast over the jam itself, placing the fly a few feet upstream in front of the jam, thereby permitting the current to collapse the leader against the concave face and the fly to eddy near the edge as though

149

unattached to any leader at all. The artificial will remain there as long as any natural and when the outlying currents begin their persuasion to carry it away, lift the coils of the leader from the debris slowly and carefully until it is entirely free, then a sharp flick of the rod tip will lift the fly cleanly and quietly for another cast. If the leader should catch in the debris, draw the line with the left hand and extend the rod up forward with the right until it meets the entanglement, then work the leader free with the rod point. If this is carefully done, it will not alarm the trout.

These are the methods to be used where the artificial hook dams have been installed, and they become increasingly important because every one of the dams is designed to hold good fish, who take the sulphur duns that glide and eddy against the concave edge exactly as they take in the natural jams. A small but important detail of the hook dam construction is to place the log covers so that they are partly submerged on the concave or upstream side; otherwise the natural or artificial will pass underneath the cover into the dam itself instead of gliding along the edge.

Now you may cast and float your fly successfully and you may raise and hook your fish, but getting them out of the undercut jams is another story. Fox claims that the only way to do it is to slack the line immediately, then kick them out of the jam from the downstream side and scare them up to the clear water where the issue can be debated on better terms. I think he is right; I know he got a three-pounder and two four-pounders that way, though I shall never forget the evening when he tried that trick with a big one, who got his directions mixed and plunged his way through the log jam on the downstream side. Fox was unable to thread his rod through the debris to follow the fish so, while the fish rested

quietly behind the jam, he quickly unhitched his leader, drew it through the debris and fastened it to his line again to continue the contest, whereupon the big fish, perverted being that he was, blasted his way upstream through the jam again, and showed Fox how a leader should be properly severed from a line.

Throughout this discussion I have used the terms "sulphur" and "blue-winged sulphur" to describe these particular mayflies. These terms actually describe the color of the bodies which are a sulphury yellow color in various shades of light and dark. All of them belong to the great genus Ephemerella, and they are close kinsmen to the Hendrickson. They are members of the same general group of light-colored flies or pale wateries in which other genera might well be classed such as the Stenonema, Ithaca, and Canadensis, betimes called light Cahills. The latter are often present on limestone waters, and I have found another, the Heterotarsale species, in goodly quantity on the northern limestone waters. I do not attach much importance to the Cahills for these waters, never having seen them emerge in flushes or fleets or on the water in simultaneous plurality, though individually they ride the water for a fair distance. On Cedar Run a pretty mayfly with mahogany body and dark blue wings emerges with the sulphurs but they burst through the surface film with remarkable ease and continue upward into the air without pausing for an instant to ride the currents. All one afternoon I tried to make a collection of them and succeeded in getting only four specimens, so elusive were they, and I finally threw these specimens away in disgust, with the conviction that the trout could not do any better than I. Let them remain unnamed and unsung!

I regret exceedingly that I cannot name the various sulphurs as to species. For two years I made collections

151

of them but had the worst kind of luck with the male specimens. Without a male spinner in good condition, a taxonomist cannot make a determination. The male spinners are rarely on the water and the few male duns that were collected failed to molt satisfactorily in the cages. Nevertheless, I expect to try again and hope to acquaint the reader with the correct determination as to species.

Apart from the formality of names, the sulphurs may be classified as blue-winged and pale-winged, the latter being a bluish-white sometimes tinged with yellow as with E. Dorothea. The blue-winged variety are the largest and yellowest in the body. They are the earliest to emerge all over the limestone country. The blue color of the wing is not nearly as dark as that of the Hendrickson but is a soft sky blue, often with just a hint of lavender. On Cedar Run the advancing season is not marked by a definite starting and stopping of the various species as they emerge at their appointed periods. There seems to be a blending of one emergence period into another, so that the latter half of the season one is not conscious of the presence of a new and smaller species until the trout begin to splash and refuse the big blue-winged imitation. Then it is time to fish a smaller fly, two-thirds the size of the earlier flies. For these reasons it is a good plan to tie the sulphurs in two sizes, 16's and 20's, the smaller preferred with a bluish-white wing or a whitish wing that flashes a good deal of light.

SPINNER

The sulphur duns have always afforded excellent daytime fly-fishing, and like the Hendrickson they also excel as spinners for some first-rate evening fishing. Along the banks of Cedar Run or other limestone streams, almost any evening in May and June, can be seen that faithful group of fly-fishing addicts, separately crouched and waiting patiently, their forms dimly visible through the screening foliage. Rods are mounted and resting nearby and perhaps a leader point is being sucked and softened for instant jointure with the fly, while necks are craned and eyes lifted skyward searching, ever searching for the first sign of the dancing spinners.

How gladdening to see the first of them, debouching from the treetops, when the hour is not yet seven of the clock. Then a hand steals to the old familiar grasp of a treasured rod, takes it and flexes it speculatively in anticipation of the crowded moments ahead. More spinners join the first group, and yet more, until a great cloud of them is collected overhead; dancing, whirling, swooping, crossing and recrossing in a raging blizzard of amorous forms. Lower and lower they drop and finally the first few begin to trade back and forth, forth and back, over the stream, yellow egg sac cradled in typical fashion and wings beating their steady rhythm. Shortly then, the greater company of spinners descend in a mass and all together they sweep the stream, upstream and down, covering their chosen beats, until they are relieved of their precious cargo. By now, the crouched and stilled forms have crept forward, rod advanced and fly hanging from the leader point but stopping a respectful distance from the furtive trout. The great finale is near, and it needs a steady hand and a cool head coupled with a clear plan of action to avoid being demoralized during the impending confusion of events.

These are a few breathless moments for the fly-fisherman until at last the first of the exhausted spinners strikes the water and a small trout splashes excitedly. More spinners fall to the water and several small trout begin to feed, and yet the watching angler moves not. Suddenly the great mass of spinners are hanging close to the sur-

153

face, compressed in a low, thick cloud, and the surface of the water becomes alive with twitching wings and bodies. Then there is heard a deep, thumping sound and a thin spurt of water flashes upward in the evening gloom; a great trout is at work. Now the angler comes to life, moves quickly to the rear for the proper angle, and loosens the first preliminary casts to get the range.

His plan is now clear—to fish for the biggest and heaviest trout that he can find, for there will be little daylight left to do more than that and if he gets into him it will probably take half the evening to bring him to net. The small ones should be ignored lest the commotion of taking one of them make the big fellow shy and keep him in his lair.

Briefly dramatized, this is the fundamental and correct procedure to be used when fishing to a heavy fall of spinners.

The first few trout, usually small, to begin the rise always provide a strong incentive to start fishing, but this should be resisted, at least for a short time after the heavy fall strikes the water. Then if it appears that the larger fish will not feed that evening the angler may entertain himself with smaller ones, until another evening when he will surely meet a big one.

At all events a company of good anglers should fish in this way, being morally bound not to stomp the banks and flail the water in advance of the fall, for if they do the better fish become timid and unresponsive.

The basic principles of fishing to the sulphur spinners are the same as those offered for the Hendrickson. It is a good scheme to get your fly to a big fish as soon as he begins to feed, since he is more likely to see your fly at the beginning of the fall than later when he is plunging about and you are trying to bracket his position to intercept his attention. During the short lull when he stops

154

and goes down to swallow his mouthful do not leave him, but be ready to cast the instant he begins to feed again.

If, when the last spinner has been washed away and the last mouthful has been taken, you have not been able to lure him to your fly, do not feel bad, for you must know that your artificial has been competing with thousands of the naturals for his favor. Look for him the next evening and the next if need be. The big blue-winged sulphur spinner brings up some of the best fish in the stream. One such, a magnificent trout, I pursued for a number of evenings; when finally I got into him, he simply lay there like a log and would not move. I could feel a vibrant thrumming of the line but no more, and then the hook came away from its hold. I wonder if that trout knew that he had been hooked.

The log jams aforementioned or piles of dead weed on other streams are hazardous affairs for the evening fisherman, though with the smaller fish a drastic and decisive mode of action can be used to overcome them.

One evening, there were at least six of the smaller fish close to the concave edge of one of these jams feeding to a rather light fall of spinners, and it was more or less easy to get their attention with the artificial. A fish that rose and was hooked could whirl and dive quickly into the debris without any difficulty. Beginning with the nearest fish, I cast and watched for the flicker of the little plume of water that is jetted with the rise. I tightened quickly and without slackening continued to increase the pressure, risking all to bring the trout around the end of the jam, through the narrow but clear channel near my bank, and into the clear stream below, where he was easily netted. Five of them I got that way, and one of them was the strangest looking trout I have ever seen. He was as deep as he was long, with a high, humped shoulder, very much like a sunfish or bluegill. Probably

155

his appearance was due to some deformity, but withal he was a strong, handsome fish.

One of the strangest peculiarities about the sulphur hatch is the fact that the hour of the spinner fall is progressively later every evening as the season advances. Anyone who is moderately observant and who follows the fall, evening after evening, for five or six weeks, is bound to discover this, simply because he has less and less fishing time with the fall on each succeeding evening. In early May the fall is likely to begin around seven o'clock, but early in June the fall may take place around nine o'clock and allow only ten or fifteen minutes of daylight fishing. This is a rough estimate of the difference and it may vary a little in the regular progression from day to day according to the state of the weather. A sudden squall during the nuptial dance will sometimes drive all of the spinners back to the shelter of the trees, from which they will not emerge again until the weather has cleared. If the weather clears just a few minutes before or after dark, there may be no daylight fishing at all when the spinners come down, and the angler may be forced to cast in total darkness, judging the accuracy of his casts by noting the direction of the sound of the feeding fish. Or perhaps he can assume a position where a moonbeam, falling across the water, allows him to see the ring of a rising trout.

Night fishing with the dry fly is never a satisfactory affair on streams. The angler is ever at the mercy of crosslines of drift and the inevitable hazard of drag, and I have always felt that such fishing was not legitimate dry-fly practice. Though I have never deliberately sought night dry-fly fishing, there have been occasions when the evening fall of spinners was delayed until dark and I felt a natural reluctance to leave the stream without making a few casts even though the artificial could not be seen on the water. At such times I have used an unusual but effective system to obtain the greatest accuracy in placing the artificial in the pitch darkness. The plan, which may be of some interest and usefulness to others, is outlined as follows: when it is plain that the spinners will be late in falling and while there is yet some daylight remaining, choose a place where a good fish has been rising on previous evenings, crawl as close as possible to his feeding position, keeping low and out of sight, then lengthen the line and make a few practice casts to the exact location where he is expected to rise. Having thus determined the exact range and location, press the reel line against the rod with the forefinger of the rod hand to keep the line from lengthening or shortening any further and hold the end of the leader near the fly in the other hand ready to be released when it is time to make a serious cast. This position should be maintained until the prospective trout is heard to feed, then the angler need only release the fly and cast the line, which has already been accurately measured for the proper distance and direction. The pick-up for each succeeding cast should be made slowly and carefully, for one never knows when the fly has been taken and there is always the risk of smashing the gut if the pick-up is made violently. It is amazing to me now to think back and recall the goodly number of times that I have succeeded in getting at least one trout in an evening in this fashion.

Sometimes it is necessary to stand or kneel in the stream to obtain a clear back cast and short forward cast. It is always best if the cast can be made with only the leader and a few feet of line thus avoiding the risk of entanglement in the darkness. All of this is possible if the approach is made carefully and quietly and no movement or ripple is made during the waiting period. Once while I was waiting in this manner a trout rose just three

157

feet in front of me, a totally unexpected thing for I had planned to fish for one some twenty feet away. I got him by simply raising the rod tip very high and allowing the fly to slip from my left hand; it was immediately wafted in front of him and he took it without hesitation.

Along with the fact that the spinners fall a little later each evening, it is also true that the duns of the sulphurs emerge later and later with each passing day. With the coming of June the great majority of them are no longer day-hatching insects. As a matter of fact, they seem to hold off the hour of emergence until the time when the spinners begin to fall; then there appears to be a simultaneous mass of duns rising upward and spinners falling downward, crossing each other vertically. It is a confusing thing to the angler, who never knows whether he is fishing to duns or spinners until the duns disappear and only the spinners remain on the water.

For a long time I believed that the hatch of sulphurs was completely finished by mid-June, but now I am convinced that it continues all summer long, but only at night. A few years ago, Mahlon Robb and I spent a fishless day on the Letort in mid-July and we remained on the stream until long after dark in the hopes of interesting one of the big night-cruising trout that are sometimes seen making a huge furrow in the water in their search for food. It was one of those quiet nights when the silence seems oppressive all around the solitary angler and nothing can be heard except the "rissing" noise of the pick-up and the occasional buzz of a terrestrial insect. In the midst of this deep silence I thought that I could hear a thin, whining sound in front of me, and it grew stronger as I strained to hear better. I could see nothing in the darkness and had no flashlight at the time but I called to Robbie, who came and played his flash in the direction that I indicated. The powerful beam re-

158

vealed a remarkable sight. Wherever it was moved over the water one could see a great mass of moving spinners of the sulphurs, as great as any I have ever seen, and I feel sure that this is the explanation for the "accidental" fall of spinners that is sometimes reported seen in the evening light at the end of the season in July or August. They are simply premature, daylight falls of the night-habit spinners, which have transformed from night-hatching duns.

Chapter 6

GREEN DRAKE

GREEN DRAKE, Shad Fly, Shaddie, and Mayfly are the common names for the giant Ephemerid, E. Guttulata of the northern limestone and freestone waters.

Near the end of May a great outcry is heard; it travels to all parts of the land by telephone, by mail, by telegraph, by word of mouth, and perhaps by mental telepathy, and it reaches every fly-fisherman. One and all they turn to listen, no matter how disposed—in quarrel or in peace, in sickness or in health, in poverty or in wealth, so great is the Circean allure of the MAYFLY.

By rail, by motor, by airplane, and on foot the great trek begins, from every hamlet and village, town and city, converging on the big northern limestone streams—

161

GREEN DRAKE

Penn's Creek, Spring Creek, Spruce Creek, Fishing Creek—from all directions north, south, east, and west, drawn irresistibly by this Pied Piper of the insect world. Anything fabricated out of the imagination would not surpass the fantasy of Green Drake time. One of the largest of its kind, numerous as the snowflakes in a blizzard, it is the lure to stir the venerable species, Salmo Goliath and Salmo Colossus, those patriarchs of the underwater world, not ordinarily aroused from their boredom or the dignity of old age.

There is no gainsaying the power of this insect to do these things; and if sometimes the rise of trout is not as good as expected, it is worth the journey to see one of the grandest sights in nature, the migration of the Green Drake spinners. Many are the incredible tales that are told about this mayfly—of dead and dying insects filling drainage ditches to a depth of a foot or more; of roads made impassable and motor traffic halted for lack of vision; of locomotives and trains weighing thousands of tons, halted and stilled; of bushes and the limbs of trees covered with a blanket of insects like a soft, clinging snow—and all of them are true.

From time immemorial its habitat has been confined to the north central limestone region, and it can be found on many of the freestone waters and on some lakes, but never has it been known to exist on the south central streams in the vicinity of Harrisburg. Fishermen thereabouts have had little acquaintance with this mayfly except that which has been gained from short annual visits to the northern streams. In recent years there has been not only a serious attempt to acquire a better fishing knowledge of this insect but an even more serious effort to bring it to these waters—The Letort, Big Spring, Yellow Breeches, and Cedar Run—by making annual plantings of the fertilized eggs obtained in the north country.

162

The history of this effort has been one of mixed success and failure, the latter inevitable where ignorance and error attended the collection and planting, the former where correctives were applied and perseverance brought its reward.

On May 12, 1948, the first issue of these plantings was seen to emerge and was captured on the Letort. In the same season they emerged on Cedar Run, ending a life cycle of two years for these particular specimens, since these two streams were the first to be stocked with the ova in 1946. An earlier attempt to plant on the Letort alone with live duns resulted in failure, for all of them died or were eaten by the birds, according to the evidence of observers.

The rare distinction of being the first to capture live specimens grown in these waters fell to my lot, the one on the Letort and the other on Cedar Run, and it might be of interest to give a brief account of both occasions since it is an historical event of no small import, at least to local anglers. Strangely enough, on the day that the first was taken (May 12) most of the same group who were party to the plantings were present or nearby, and no more fitting reception committee could have been arranged to welcome the first arrival.

All of us were widely separated, occupied with our individual fishing problems, and I was stationed in the meadow below the fishing hut on Fox's water, where the stream makes two sharp bends in the form of a letter S. While I was standing there, idly watching the stream, a movement on the surface, in the middle of the S stretch twenty-five yards away, caught my eye and I moved toward it expecting that it might be a rise. While I was still a considerable distance away, the movement recurred and then a living fluttering creature rose from the surface in laborious flight toward a bank of trees a hundred feet

163

away. It had hardly begun when a suspicion of the truth flashed in my mind and I was immediately under way, running full tilt toward the bank of trees and dropping all of my paraphernalia and trying desperately to unhinge my insect net. We reached the bank of trees neck and neck, and I made a wild sweep of the net when the creature was about to enter the foliage, quickly throwing my left arm over the mouth to prevent escape, though I was not sure that I had captured anything. Breathless and somewhat excited I moved to a clear space in the meadow and slowly opened the net to discover that I had got the first Green Drake that has been known to come from the Letort.

The Cedar Run specimen was taken under more colorful circumstances, as follows: on the evening of May 15, 1948, I was fishing with Fox and Mahlon Robb at Cedar Run some two hundred yards below the point where we had originally planted the eggs of the Green Drake. Mahlon and I were fishing together, Fox having remained upstream somewhere. At this particular part of the water the evening fisherman looking upstream from the left bank is always blinded by the setting sun, so that it is very difficult to see a rise or a fly on the water. Mahlon had been telling me about two very good fish that he had seen feeding there on sulphurs several evenings before, and while he was trying to describe their exact location·to me, both of them rose separately in quick succession. He generously urged me to try for them. I went below and threw a lucky cast to the nearest fish, who promptly rose and was hooked and duly landed. I had no such immediate luck with the second fish and needed Mahlon's help to direct my casts against the blinding glare from the water. About the fourth or fifth cast the second trout rose, was hooked, and tore downstream and around a sharp bend below me. I followed

164

and held him, then he made a run upstream and stopped to play deep for a while. Suddenly he tore downstream again and came thrashing to the surface, while at the same time my rod point was jerked behind me and I turned to see my line running upstream.

Completely flabbergasted by this queer turn of events, I could only stand for a moment looking alternately below at the thrashing fish and above at my line, pointing upstream. I decided to follow the line rather than the fish and walked up the stream bed until the line emerged from the water near my feet. I kicked around for a moment, felt something move, then reaching into the water closed my hand on a hard, flat object. I brought it up to the light of day to discover that it was a barrel hoop encircling my line. While I passed my line and rod through the hoop to free them, Robb's peals of laughter were ringing up and down the glen, and he promptly left me to tell Fox about my circus fish, vowing at the same time that it should receive top billing at the next meeting of the Fly-Fishers' Club. Though I felt a little silly about it all and could expect some good-natured chaffing, there was no denying the humor of the situation.

Hardly had Robb left me when, magically, before my startled eyes, not more than twenty feet away, a pair of huge greenish yellow wings popped to the surface of the stream, twitched and rocked for a moment, and then fluttered diagonally across-stream toward me and stopped at my feet. I stooped quickly and gently closed my hand on the big fluttery thing. There was no doubt about it. It was the Green Drake, a big female, and I put it in my collecting bottle with a great deal of satisfaction.

Shortly thereafter Fox came along the path and without delay began to advise me of the position of a very fine trout that he had seen rise some evenings before and wanted me to try for, at the same time pointing out the

165

same location that Robb had showed me. Evidently Robb had not seen or talked to him in the interim. Feeling that I was entitled to enjoy a little joke of my own, I let him complete his directions then told him quite offhandedly that I had already got that fish. Fox seemed nonplussed for a moment then immediately launched into another course of advice about another very good fish that I should try for and then pointed out the exact location of what had been the circus fish. In the same careless attitude as before, I told him that I had that fish too. Taken aback by this information, he recovered and demanded in strong terms, to know, what else I had got. I told him that I had a very pretty lady with a green complexion that he should meet and forthwith drew the collecting bottle from my pocket. I must say that he was properly impressed and quite as elated as I.

To paraphrase the remark about the swallow, one Green Drake does not make a hatch, though some were reported by other anglers who frequent the streams where they were planted. A good yearly hatch is not built up in a year or two, not unless it is possible to make plantings in the same magnitude as that effected by nature, though I am convinced that a "fishing hatch" could be quickly created if only there are enough willing hands and the best equipment to make heavy collections of the eggs.

Even with the heavy plantings there are many hazards to the eggs during the period of gestation. For several years our plantings were made by anchoring long, narrow, egg-bearing strips of cloth in the stream bed, one end fastened to a stake and the other allowed to wave and move constantly with the current. We fondly believed that they were entirely safe from foraging creatures, who could not deal with the moving strips. No one thought to question the method until 1948, when we dis-

166

covered that the cloth strips in Cedar Run were entirely covered with a mass of scavenging shrimp, more plentiful in this stream than in any other that I know.

It was a dismaying discovery and certainly it must account for great losses among the eggs. A better method is needed to overcome this risk; the best that I can conceive up to the present time is the construction of small wire cages, with mesh so fine that the smallest adult shrimp cannot enter but large enough to permit the hatched larva to drop to the stream bed. The newly hatched larva of the Green Drake is of microscopic size and does not need a very large opening to escape to the stream bed, where he digs in without delay and is thereafter completely safe from his possible enemies.

What the future holds in the way of good hatches for these waters is hard to say, though the initial appearance of the comparatively few Drakes ought to prove that ecological and climatological factors are suitable. There is every reason to believe that they should thrive, since they exist to the south of this area as well as to the north. Big Hunting Creek in Maryland, for example, has a magnificent yearly hatch of E. Guttulata, and I doubt not that it can be found on other southerly waters. In any event it ought to be encouraged to thrive by all conservation-minded people for its economic value and recreational advantages; desirable too because it is a harmless creature, not dangerous to foliage or crops since its mouth parts are atrophied in the winged stage and it can eat nothing. As a larva or nymph it lives only on decayed vegetable matter and other herbaceous items found on the stream bed.

Desirable as this insect may be, and though it provides the most spectacular kind of fishing at times, there is an inclination among many anglers to rate it too high in comparison with other species of mayflies. I have never

167

seen it more often or pursued it more diligently than during the years when I have been a party to the annual collections of eggs for transplanting. During those years I was always on hand where the best and heaviest hatches took place, since we were advised by telephone or telegraph by scouts on the best streams, and I was always ready to fish as well as collect; and I am struck, upon reflection, with the realization that the fishing has been exceedingly poor with this insect.

In recent years it has become a common thing to see a great army of anglers standing idle on the banks, now on one foot, now on the other, regarding morosely the great mass of moving spinners and each other while not a single decent trout is feeding. With the duns it has been even worse, though fishing to the duns has never been as good as fishing to the spinners. I suspect that much of this inactivity is really due to the scarcity of trout, something that no one seems to want to admit. Not that the streams are entirely devoid of trout, but to have good fly fishing there must always be enough trout in a stream so that at any given time a certain number of trout are ready and willing to feed. All trout do not feed at the same time, as we should suspect. Not even human beings, with their 8–12–6 o'clock lives eat that way.

The scarcity of trout, a grim reality now, is unavoidable with the enormous increase of fishermen in late years, and there is no Fish Commission or any other Federal or State agency that can possibly meet this demand with expensive artificial stockings of hatchery trout. Certainly it is nothing like it was in former years, when huge trout, reckoned in pounds, were taken on all of the Green Drake waters. Some wonderful tales can be told by the northern fishermen, who used to follow the hatch from beginning to end, from stream to stream as

168

the Drake emerged on successive dates, to enjoy nearly a month of superlative fly-fishing.

It is a distinctive and much more extensive application of the principle of following the hatches than is understood elsewhere. From long association with the Green Drake these anglers of the north central district have discovered that the developing hatch on any of these streams is a progressive affair which reaches a peak on different sections of the stream, traveling upstream with each passing day; and the knowing angler moves with the peak in order to be on hand where the hatch is heaviest for each day of the entire emergence period.

Beginning at the lowest stretch of the water where the Drake is known to emerge, each successive climax may take place one-half to one mile further upstream on successive days for a period of perhaps one week or ten days. According to the best opinion, this is largely the result of a difference in water temperatures, the colder water above delaying the emergence of the duns somewhat so that the hatch is spread over a considerable period of time. It is an interesting theory and may have some factual basis, since there is some evidence that the hatch as a whole may be retarded or advanced by a variation in the yearly average temperature; but there is some reason to doubt its correctness, since it has not been shown that the last Drakes to emerge upstream have a longer nymphal cycle than the downstream Drakes. Even though the former emerge a week later than the latter it must be remembered that they were planted a week later by the spinners in some previous season, so that, actually, all of them would have the same yearly or biennial or triennial life cycle.

Perhaps a better reason for this progressive upstream emergence schedule is an historical rather than an environmental one. The Green Drake is an ancient inhabitant of this earth, having been present in a slightly different form in the dim prehistoric past. The migrating habit of the spinners is a wise provision of nature which insures the perpetuation of the hatch in a given locality, and if the spinners did not fly upstream a little way before oviposition, the eggs which are planted each season might drop lower and lower downstream with the movement of the flowing current until every one of the mayflies was washed out of every stream in the land. This same migrating habit, perhaps more pronounced in an earlier day, might easily account for the progressive emergence schedule, since the immigrant spinners would have come from lower waters, always traveling upstream and continuing to establish new colonies during the passing centuries.

All of this is a matter of speculation; one theory is as good as another. Besides, it is extremely difficult to establish anything definite about the life cycle of this mayfly since it is known that individuals of the same brood may emerge in one, two, or three years. The reason for this is not known. It is generally accepted that the majority of any given brood have a nymphal period of three years before emergence as duns, and this is probably true; but we now have positive evidence that many of them issue at the end of a two-year nymphal period, according to our findings on the newly planted waters of the Letort and Cedar Run. It is likely that the relative availability of food and oxygen and the metabolism of individual species may have some bearing on this question.

Whatever the reasons may be for these peculiarities of the hatch, it is a fact that they exist to provide good flyfishing for those who know how to follow the peak of emergence. Even though of short duration on any stream, perhaps a week or so, the time of emergence varies

170

enough on different waters so that when the peak has been followed completely on one stream, the angler may then transfer to another where the Drake emerges at a later date and pursue the same routine. In this manner fishing to the dun or spinner is a daily possibility over a spread of three or four weeks for persisent, migratory anglers, particularly for those who live in the vicinity of that group of fine limestone streams in the north central counties.

One of the strange results of the plantings on Cedar Run and the Letort is the fact that they emerged in the second week of May, whereas the eggs from which they issued were collected from the ovipositing females of Honey Creek on June 11 two years before and planted the same day. It means that the emergence date for the same breed of mayflies has been established a month earlier than it occurs on Honey Creek, sixty or seventy miles away. This is the earliest emergence date of which I have any knowledge, and indicates the possibility that fishing to the Green Drake in the future may encompass an even greater spread than before. It is an exciting prospect, but, the idea must be tempered with a little caution simply because the rise of trout is not always what is should be. Even in those earlier days when fishing was generally very good, the angler had to depend on the evening fall of spinners for the best results.

The dun or subimago is a big creature which, standing alone, is bound to draw the angler's attention since it overshadows a dozen of the obscure sulphurs surrounding it, although the latter are, nevertheless, being taken by the trout. Very often it happens that way; it is not unusual for a whole company of anglers to disregard the duns and switch to an imitation of the sulphur. On Honey Creek I have seen the trout ignore the subimago of E. Guttulata and take the dun of E. Simulans, the

171

brown mayfly, a close relative of the Green Drake. With me the dun fishing has been so disappointing that I like to depend almost entirely on the spinners and the imitations of the spinners.

I am inordinately proud of the present imitation that I use—the lightest in weight, the best floater, and the easiest-to-cast big fly that I have ever seen. It is the result of the discovery of the porcupine quill, hollow, buoyant, and durable, joined with a comparatively small short-shanked hook to keep it light in weight. Moreover, the novel method of applying hackle fibers makes a fine wing, big as one cares to make; durable and easy to dry, yet maintaining its shape; and avoiding the common fault of big mayfly wings—the tendency to whirl or spin in casting. I shall describe the wings only briefly since they are detailed in the more specific directions in another chapter. They are made by first fashioning a densely fibred palmer fly turned on the entire length of the thorax *only*. Then the hackle is clipped away above and below to leave a flat, broad plane of hackle fibres flaring away on either side of the thorax, each fibre separate and independent of the other and retaining its position by self-recovery. Of course it is a comparatively recent pattern, not having been tested in those years when the Green Drake fishing was at its best. No matter, its small history of successes is enough to establish it as truly worth while.

The fishing to the spinners of this large mayfly is very much like that which is practiced for the Hendricksons and the sulphurs. But there is one notable exception that ought to be mentioned, namely, the fact that lines of drift need not be strictly observed. It is true that fish can be found in the stronger currents feeding on the drifting spinners, but the big fish, the giants of the stream, are usually outside the drifts, not a strange thing at all, because the fall of the spinners is ordinarily so heavy that there are just as many falling near their resting places as there are anywhere else. They need not move to a feeding lane to get their share. Whether in or out of a line of drift, choose to watch a place where a good fish should be resting—a place with depth and good cover. If it is a log, a stone, a cut bank, or a brush heap that he uses for a home, he is likely to be found feeding within inches of it.

In the earlier chapter on the Hendrickson some attempt was made to acquaint the reader with the peculiar rhythmic manner in which a trout feeds during the progress of the hatch; how the regularity of its rises increases to a rapid rate; how it falls away to a slow tempo when a considerable number of insects have been taken, and how finally, when the trout is completely gorged, there are but few rises at long intervals. This is one of the tremendous secrets of dry-fly fishing; no one can be an accomplished fly-fisherman until he is thoroughly familiar with this aspect of the trout's behavior. And yet, strange to relate, it is rarely mentioned or exploited, perhaps never really observed. Even so I must confess that it does not seem to apply to the dun fishing with the Green Drake as strictly as it does to the dun fishing with the lesser insects. By that, I do not mean that the same rhythmic manner of feeding is absent, but it seems that I never find trout feeding on Green Drake duns except in the very last stage, when rises are few and a long time apart—as much as fifteen or twenty minutes. Very often when I have taken a trout on the dun pattern I am astonished to discover that his stomach is charged and distended like a country sausage; when autopsied the entire contents are delivered whole, so closely are the insect forms jammed and packed together. It is plain, in such cases, that Sir Trout was a very busy citizen at some

173

time or other during the hatch and must have practiced the same rhythmic manner of feeding hereinbefore mentioned. The only sound reason that this activity goes unnoticed by me or by others is the plain fact that Green Drakes are big meaty fellows on which a trout could be quickly filled in a few minutes' feeding time, therefore arriving at the last stage of his feeding period at a comparatively rapid rate.

If any success at all is expected in these difficult circumstances, it can be gained only by an extended period of watching and waiting and timing the cast precisely to meet the long-awaited rise. This means that if the angler has spotted a rise he must wait for the next succeeding rise to determine the length of interval between the rises of that particular fish. They may be as much as twenty minutes apart and that means another fifteen or twenty minutes' wait before beginning to cast, or a total of forty or fifty minutes for one fish; but it can be much, much longer if he chooses a natural instead of your artificial.

Once I spent something like two hours to get one fish, a good one. It happened on Penn's Creek, famous for its Green Drake fishing and my favorite stream for this kind of fishing—big water, big trout, and a fine hatch regularly.

Fox and I went to the stream last June, established ourselves at a local hotel, met old friends and acquaintances, then went to the stream in midafternoon and were confronted with the most stupendous hatch of duns we had ever seen. Like everyone else we marveled at the fine hatch and the curious lack of fish showing anywhere. Most fishermen in these circumstances like to move about a great deal searching for feeding fish and never really stay long enough in one spot to see a rise. Thoroughly chastened by the sorry results of this practice in former years, I moved about as little as possible while

174

many anglers passed me by during that long June afternoon, and finally I saw a thumping good rise in midstream, some sixty or seventy feet away. The stream is extremely broad and flat in this area, with a featureless surface, making it rather difficult to spot the location of rises very accurately, so I moved along the bank until I had the approximate location of the rise aligned with a tree on the opposite bank and then sat down to wait for the rise to be repeated. A period of twenty minutes must have elapsed before the rise occurred again, and I began the slow process of wading out very carefully to a casting position some forty feet from the expected rise. When a period of some fifteen minutes had gone by, I began to cast as carefully and accurately as I could judge the location of the trout. Accuracy at this stage of a trout's feeding period is of the utmost importance. A cast which is off his feeding lane to one side or the other won't tempt him, and the margin is only a matter of inches. Imagine my chagrin when the expected rise took place to a natural and I discovered that my cast was short by two feet at least. There was nothing to do but lengthen the cast immediately to the exact spot, lock the reel to prevent shortening or further lengthening, allow the line to trail downstream and wait another fifteen or twenty minutes. When he rose the next time he took a natural instead of my artificial—poor timing on my part. I had managed the whole thing very badly but I had no other prospects so I waited a while longer in the same spot and finally got into him. We had a furious time of it for the next twenty minutes. Penn's Creek is big, rough water with heavily wooded banks, and a fisherman must stay in the stream with his fish. It needs some artful wading, a heavy hand, and strong tackle to manage a good fish in such water but finally I got him out of the current and into the quiet water and netted him.

He was a strong, heavy fish, too beautiful to kill, but I was overwhelmingly curious to see what his stomach would reveal, not merely in kind but in the all-important matter of quantity, in order to confirm or deny my estimate of his appetite and his feeding pace. The autopsy disclosed what I had expected, a stomach completely loaded with Green Drake duns, and hardly any room for another.

As an incidental matter to the fishing on these waters, I shall mention a peculiar method, sometimes called "bouquet" fishing, practiced by the local residents of these parts. I suppose that it has its counterpart in the blow-line fishing with live duns which prevails in other parts. As I understand blow-line fishing, only one live dun is impaled on a small hook, attached to a light silk line which is allowed to "dap" on the surface until it is taken by a trout. Possibly the more self-righteous fly-fishermen would regard this method with some tolerance, but the local residents do not admit to such refinements. Not one but many live duns, as many as a big No. 4 or 6 hook will hold, are impaled, wriggling and squirming, until the whole affair looks like an exotic greenish-yellow bloom such as no man or beast ever saw; and then this whole assembly is slung without fear or favor with a most ungraceful heave in the general direction of the water, where it is allowed to sink and go where it will. From my observation the "bouquet" is seldom taken by a trout. It is a distasteful affair from start to finish, including the business of constantly beating the bushes along the banks in order to make a collection of the live duns.

Earlier in the evening, before the female begins to fall, there is often an opportunity to fish to the male spinner, the erstwhile coffin fly, and the only male spinner of any mayfly species on limestone waters that is worth imitating. For a little while before the females leave the shelter of trees and foliage, the males can be seen over the water as much as on the land, swooping and darting through the air in search of a mate, sometimes landing on the water and remaining there for a few moments to rest before rising again to continue the search. When the romantic play of the sexes is over and the fertile females begin their mass migration upstream, many of the males fall to the water and float downstream to become the first attraction to the trout, but never in any quantity like the females. The fall of males lasts for a comparatively short time; then there is a definite lull, when few males can be seen, until the first of the females begin to fall and the fishing begins in earnest.

On those days when the Fly-Fishers' Club planned a collection of eggs, it was regarded as a cardinal sin to bring along any fishing tackle lest it should cause a distraction so powerful that it would keep anyone from performing his bounden duty. I confess that I sinned many times, secreting my tackle in the rear of the automobile before it could be noticed and invite a hail of protests.

On reaching the scene of operations, usually around five or six o'clock in the evening, I lost no time in retrieving my tackle and waders, shamelessly and brazenly, before the staring eyes of the whole company, heedless of the scurrilous remarks aimed at my retreating figure.  In this way I managed to enjoy a moderate amount of fishing with the male spinner; when the lull came in the male fall, I knew it was time to quit and would hurry back to my post and seize my big insect net, determined to outnet, outnumber, and outlast everyone else to salve my stricken conscience. I think I have been forgiven.

For all of us, it was always a terrible sensation to stand in the middle of the stream, swinging the nets against the oncoming hordes of fertile spinners while

177

here and there the distinct plop of a feeding fish could be heard. And added to this frustrating situation there were always the jibes of the local gentry, standing on the banks and marveling hugely at this strange spectacle. On at least a dozen such occasions I kept my rod mounted and ready near my netting position, but though I looked longingly at it many times, never once did I weaken. I swear it.

It was interesting to learn that the bodies of the male spinner and the female after oviposition are exactly alike in color. Many examinations of thousands of females in the collecting tubs showed that they had the same chalky white appearance as the males. The eggs are quickly extruded when they are cast into the tubs; in fact, the spinner has no power to halt this extrusion. In every case the female was filled with a gas which made a tiny explosion when the empty tube was squeezed. There is no doubt that this is manufactured by the spinners as an aid to the undulations of the body in expelling the eggs, which issue in two columns, from double oviducts.

The difference in color between a loaded female and a barren one is so noticeable that a seasoned netter can distinguish one from another while they are passing him, the one being a buttery yellow and the other a chalky white, and he can even distinguish a half-barren female, though it needs keen eyesight to do this at the height of the migration. For these reasons it is fair to contend that many standard imitations are incorrectly tied when they have the cream or yellow bodies that one often sees.

The ultimate position of both sexes on the water is the traditional prone position, i.e., wings extended on either side of the thorax, flush with the surface. The bulk of the males are taken in this position but not the females. For the latter I prefer the intermediate position with the wings partly elevated, for that is the way the trout get

178

them at the beginning of the fall and during the time of the greatest feeding activity. The great mass of female spinners do not become prone until after dark, when it is no longer possible to fish accurately. It seems to be an important distinction to the trout.

On the entomological side it might interest many to know that the Green Drake of all these streams is the species E. Guttulata. Wetzel established the Drake of Spring Creek as being of this species and so are those from Fishing Creek, Spruce Creek, Honey Creek, the Kishacoquilas, Stone Creek, and now the Letort and Cedar Run. Specimens from these various sources were collected and sent to B. D. Burks, associate taxonomist of the Illinois State Natural History Survey, who made the determinations. Though all of these are identical species, there is a marked difference in the color of specimens from different streams; again we are confronted with nonuniformity. On Spring Creek the Drake is predominantly yellow; on Spruce Creek, Honey Creek, and the Kishacoquilas of a brownish cast; on Stone Creek, the greenest of them all. I never truly understood the appellation "Green Drake" until I saw the Guttulata of Stone Creek, which is green enough to deserve the name.

Chapter 7

JAPANESE BEETLE

WHAT a rare thing it is to witness the advent of an entirely new insect on trout waters. I do not know of a single recorded instance of this nature in all of fly-fishing history. Not that the insect of which I am about to speak is new to the entomologists, but it is certainly new in fly-fishing practice and such an insect that would compel a Halford or a Lunn to abandon anything else to study, to imitate, and to fish.

Popillia Japonica is a very pretty name for a very pretty beetle, perhaps the handsomest of all the beetles— a strong composition of red-bronze, shiny green, and bronzy black, mild of manner and rather unobtrusive. No one without previous knowledge would suspect him of

181

being one of the most terrifying and devastating colonizers of our time.

Originally confined to the islands of Japan, these beetles found their way to the United States about the year 1916 when a few specimens were discovered and identified as Japanese beetles in a nursery near Riverton, New Jersey. These insects have some relatives in this country, but they themselves had never before been seen or recognized here. Entomologists are convinced that they were introduced as larvae in the soil about the roots of nursery plants, possibly azalea or Japanese iris.

After the first discovery, no more than a dozen-odd beetles could be found that year in the neighborhood of Riverton, and the area of infestation included a total area of less than one square mile. In 1917 and 1918 the spread was hardly noticeable and did not include more than a few square miles. In 1919 a forceful notice of the awful menace was served on the agriculturists when it was discovered that the beetles were alarmingly abundant in an area of some 48 square miles; in that same year it is likely that they crossed the Delaware River and reached the Pennsylvania side, where they were found for the first time in the following season. At the end of 1923 it was found that the territory infested by the beetles amounted to 2,442 square miles, covering wide areas in New Jersey and Pennsylvania, a truly astonishing increase.

In 1916 only a few beetles could be collected, but in 1919 as many as 15,000 or 20,000 beetles could be collected by hand, by one person, in a single day. During the summer of 1923 beetles were to be found by many thousands in individual orchards on the foliage and the fruit of various trees. This condition obtained not only in a single orchard but in many orchards throughout the heavily infested territory. Here is an example of how abundant the beetles were during this season: early one

182

morning, while it was cool and the adult beetles were more or less inactive, a large canvas was spread under each of 156 ten-year-old peach trees; the trees were shaken vigorously and the inactive beetles, instead of flying to other plants, as they would have done later in the day, dropped to the canvas and were collected. In a little over two hours 208 gallons of beetles, or 1⅓ gallons of beetles to each tree, were collected. On examination twenty-four hours later it was found that, in spite of the removal of the beetles during the previous day, the trees were about as heavily infested as they had been the previous morning. Besides attacking the foliage, the insects feed on the fruit and are found clustered on apples and peaches in large numbers. As many as 278 beetles have been collected on a single apple.

During the fall of 1923 larvae were found to be extremely numerous in some locations; for example, on one of the greens in a Pennsylvania golf course they were found to the number of 1,531 to the measured square yard. In pastures the infestation was found to be as high as 717 larvae to a square yard.

The adult Japanese beetle is a beautiful and brightly-colored insect, varying in length from ⁵⁄₁₆ to ⁷⁄₁₆ inch and in width from ³⁄₁₆ to ⁹⁄₃₂ inch. It is broadly oval in shape, moderately convex, and shining. The upper surface is somewhat flattened, without pubescence, hairs, or scales. The color is a bright, metallic green except for the greater part of the wing covers, which are coppery brown. The undersurface of the body is clothed with short, grayish hairs, and the legs are of a dark metallic, coppery-green color, varying in tint in different positions.

The total life cycle of the Japanese beetle is one year, five-sixths of this time being spent in the soil as an egg, larva, or pupa. Having passed the winter in the soil at a usual depth of between 2 and 4 inches, the larvae become active late in March or early in April and feed actively until they change to the prepupal stage in May and June. The prepupal stage may last for ten days or as much as two or three weeks. The larva then transforms to a pupa, and the adult emerges about two weeks later. In some cases the adults may remain in the pupal cells for several days before emerging, but on the average the adults issue from the ground on the second or third day after transformation. In the area around Riverton, New Jersey, the first beetles to emerge were usually found between June 10 and June 20. The average emergence date for females in 1921 was June 20 and for males June 21. Although the first beetles were found June 10 on the heavier soils, it has been found that the beetles emerge from a week to ten days later on the soil of the sandy types. In 1922 the first beetle was found on June 11, and the average date of emergence was June 18.

After emerging from the ground the beetles usually climb on various low-growing plants, and if the weather is clear remain for a few hours without feeding until they are thoroughly hardened.

During the morning the beetles concentrate on low-growing plants, such as smartweed and beans. As the heat increases during the day they become more active and disperse to the taller plants until early in the afternoon, when they are abundant on the tallest elms, oaks, and maples. After 3 P.M. their flight is toward the ground and the lower-growing plants.

There is a general increase in the activity of the beetles during the day until 2 or 3 P.M.; after this time their activity decreases until dark, when flight ceases.

On clear days, between 8 and 9 A.M., numerous males can be observed flying low over the ground in search of emerging females. During the early part of the day this tends to concentrate the beetles on low-growing plants.

184

In 1922 this habit was particularly noticeable on the golf course at Riverton. On clear days the beetles' flight commenced between the hours of 8 and 8:30 A.M., and the first beetles found on the wing were invariably males. As the females would issue from the ground the males would alight and attempt copulation before the females had opportunity to take flight. This resulted in a large number of males alighting on and near the female, and in their attempts at copulation the congregated beetles resembled balls. From one of these masses with only one female present 198 males have been collected. It was observed that in alighting the males always approached the female against the wind, apparently being attracted by the odor, and when alighted on the ground it was usually from 4 to 6 inches on the leeward side of the female, which they approached by crawling along the ground. As the males drew in rapidly, they formed trail several inches long extending from the ball which was forming about the female. As the wind would shift slightly the direction of the trail of males would change within two or three minutes, so that the trail of beetles was always directly opposite to the direction in which the wind was blowing. The habit of balling usually ceases about midday, not to recur again until the following morning. In an area of 25 square yards on a golf course, 78 balls of the beetles were observed at one time, each ball being composed of a single female and from 25 to nearly 200 male beetles.

It is now evident why the Japanese beetle has been so successful in establishing itself in this country. The adult beetle is a voracious creature, attacking many important crops such as the fruit and foliage of the apple, peach, cherry, grape, beans, clover, alfalfa, and sweet corn, in addition to various shade and ornamental trees, shrubs, and herbaceous plants. Since the insect is a fairly

185

good flier and is exceedingly active on warm days, its normal spread from year to year has been fairly uniform. The fewness of its natural enemies has permitted the beetle to reproduce in almost incredible numbers, while climatological, ecological, and cultural conditions in the infested areas have been remarkably suitable for its development; there is every reason to believe that this insect is now firmly established in the United States.

The foregoing account is taken from records of the United States Department of Agriculture and is a wonderfully gripping story of the beetle's invasion, more than a matter of mere statistics. Its entry heralded a titanic struggle, quietly but intensely waged on an ever-advancing front, on the one side by the incredible multiplication of this destructive insect and on the other by the energies and resources of several powerful states and the Federal Government, marshaled together in a common defensive and offensive warfare to halt the disastrous pillage. It is no less intense because it lacks fanfare.

I cannot say exactly when the encroachments of the beetles reached the limestone country, though its arrival was anticipated and feared by the agriculturists long before it appeared.

I believe that the bass fishermen of the Susquehanna were the first to call attention to the beetles' extraordinary position in the menu of the fish.* Bass fishermen from these parts used to make frequent and regular visits to the great Conowingo Dam on the lower Susquehanna, some eighty-odd miles from Harrisburg, there to experience the marvelously good bass fishing that once existed. They brought back wondrous tales of big bass,

* See *Advanced Bait Casting*, Charles K. Fox. New York: G. P. Putnam's Sons, 1950.

four- and five-pounders, sipping beetles from the surface of the river, making no more disturbance than a small dimple like the rise of a fallfish; tales of huge bass autopsied and revealing beetles crammed from mouth to vent and spilling from either end upon the slightest exertion. Naturally these stories had little or no meaning to the trout fishermen hereabouts, not until the beetles had arrived and he saw before his eyes the plain evidence of a new aspect of fly-fishing.

As nearly as I can determine, the first formal pronouncement of the new possibilities for the dry fly were made in the *Pennsylvania Angler*, angling paper without peer in the land, issue of February 1944, by Charles K. Fox, who was the editor at that time. It was a forward-looking article and offered some valuable hints on constructing and fishing this artificial. The article described an experimental tie of the beetle for which a coffee bean was used; a bisecting slot on the flat side of the bean was fitted to a hook shank, and fastened with one of the synthetic, waterproof adhesives. It held best when the bean was glued to the woolly bodies of flies with wings and hackles removed, rather than to a bare hook shank.

I have tried this construction at times; it holds together fairly well, but it is not as durable as one would wish, though it is easily and quickly made. A small triangular file and a tube of cement are all the equipment that is needed.

I am sure that I did not appreciate the beetle until about 1942 or 1943. By then, a man would have had to be completely blind and devoid of all perception if he could not see the dazzling prospect of the new fishing, evidenced by the great hordes of this insect near the limestone streams and the consequent activity of the trout. At that time there was very little, in fact no information available to the fisherman in any book or paper,

187

until the above-mentioned article. In the intervening years, only a relatively few fishermen have tried to gather an adequate understanding of this fishing. Much has been done, but something remains to be done if the whole story is to be told. Whatever information I can submit here is given with the provision that it is subject to revision or improvement according to additional information adduced in the future.

This much I can say with assurance: that the general approach to a study of the beetle must be made in the same way as for any other insect, i.e., by observing the appearances and habits of the beetle on the water and the habits of the trout in taking them. These are the guideposts upon which every fly dresser and fisherman should depend in arriving at the construction of an adequate imitation and the proper manner of fishing it; if they are followed, surely within good time a satisfactory formula will be evolved.

I have written earlier on the desirable habit that Letort trout have of feeding with constant regularity on the surface during the daytime. Much of this feeding habit is due to the presence of this beetle and other terrestrials during the heat of the day. There is no evening rise to the land-bred insects, not in the sense that one understands with the spinners of the Ephemeridae. True, there are night-hatching terrestrials such as the June bug, various moths, and the lightning bug or firefly, the last of which affords some evening fishing, but nothing like the mayfly spinners.

The best of the fishing with the beetles takes place during the hottest time of the day, midafternoon, according to my observation. These insects are often active in morning on a hot midsummer day, but the trout do not usually become interested until the sun has raised the temperature of the water. On the Letort they begin to move around eleven o'clock and are at their best about two or three o'clock, ceasing their surface feeding around five or six o'clock, though there may be an evening rise if there are mayfly spinners coming down. It is plain, then, that the fishing to terrestrials is largely governed by the heat, without which these insects remain somewhat chilled and more or less quiescent; the wind is an additional factor, though the beetle is often on the water without this influence.

The most outstanding thing about the beetle's behavior on the water is his strict obedience to lines of drift, which he observes far more rigidly than any other insect that I have seen in my career as a fly-fisherman. Once he has fallen to the surface he seems to lose all power to direct his destiny; not able to move himself an infinitesimal fraction of an inch, he exhibits no more than a slow futile movement of the legs. He is the direct antithesis of the grasshopper, that peasant of the insect world, who tosses his small pride to the winds and makes a wild scramble for the safety of the bank. With the beetle there is not even a bit of twitching with the wings, and not the least effort to lift himself from the water; and the trout, particularly the big trout, take them with a matching mood, quietly and unobtrusively.

It is a curious thing about a big trout taking beetles. Nine times out of ten he is very close to the bank, sheltered by the grass or an overhanging bush, never advertising his presence or his occupation; when he is caught it is always a great surprise to find him full of beetles, since no one would have suspected that he had been busy all afternoon collecting them. He likes to lie near a line of drift that is concentrated near the bank. bringing him many delectable gifts, though once in a while a big trout can be found in midstream if there is good cover nearby. In the main, however, it is best to watch carefully the overhanging banks and the shade or cover created by shrubbery, even though it may be necessary to sit and watch for a long time before a rise is detected with certainty.

Sometimes a trout's probable presence can be calculated by tracing a line of drift along its meandering course until it strikes a likely place near one bank or the other, and then fixing the eyes on that spot for a lengthy period, at the end of which the fortunate angler is rewarded with the sight of a rise. Fishing to the beetle needs a little patience, but the oppressive heat of midsummer often enforces a lazy attitude and the angler can use these periods of idleness to sit and look for rises.

There is one especially good spot on the Letort, one which I have mentioned before, where the drift is concentrated near the trunk of a box elder and some overhanging shrubbery. Fox and I used to take turns watching it very carefully during the midsummer heat when the beetle period was in full bloom. A passer-by might have marveled at the lack of industry on the part of two lounging fishermen, who showed no more sign of activity than a thin wisp of cigarette smoke curling lazily above their heads. Nevertheless, he would have been surprised to learn what they saw and accomplished at that one small spot. Here is the record for one brief period: I got one that was 2½ pounds and lost two that were near 3 pounds each. Fox got one of 3 pounds and lost one that was somewhere between 3 and 4 pounds.  At this same spot Fox got hold of a big one that weeded him solidly and he decided to wade into the stream "up to his cigarettes" to grub for him. James Kell, Jr., and Ross Trimmer were there at the time and offered many amusing comments on the ethics of such tactics. I have a photographic record of the whole dripping episode.

For a short time, during the height of the beetles'

190

emergence period, the grasshopper will be on the water at the same time and the trout will take both of them, though it looks to me as if they prefer the grasshopper. At such times a trout may have been taking many beetles unnoticed before revealing himself with a slashing, explosive rise form to a grasshopper that happens to come his way. If the grasshoppers are scarce that day, the angler may not see the rise-form again and may think that he has seen a "one-rise wonder," but he should not be deceived, for that same trout may have gone back to his quiet work with the beetles. It is worth a cast or two.

On most streams a fair imitation of the beetle that floats works very well. The coffee-bean imitation is good enough on the Yellow Breeches and some other streams. Various other compositions of the beetle have been tried with reasonably good results. But the worst failure I have ever had with any imitation of the insect was with the trout of the Letort. For one whole season I failed to tempt a single decent fish with any imitation that I could devise, including the coffee bean and some imitations designed by others. It got to the point where I began to carry a streamside fly-tying kit, a detestable thing to me, for I do not have a disposition which allows studious concentration for fly tying at streamside while a trout is rising repeatedly and blatantly not more than ten yards away in full view.

Nevertheless, I made many on-the-spot imitations as my fancy dictated but never a successful one by this method. If I had been alone in my failure it would not have been so bad; then I could have sought and learned from others more successful and bettered my fortunes, but others were equally baffled and no more successful than I.

It has always been like that on the Letort; there has never been an easy solution for anyone on any problem

191

concerning these trout and the insects on which they feed. I do not know why it should be so. Perhaps it is because they receive a superior education or training or perhaps a special course in similitudes and differentiation. Never have I seen such a stream and such trout; I truly believe that anyone who can take fish consistently on the Letort can take them anywhere else in the world. There is this one saving grace: once the secret is learned, success is fairly regular thereafter.

So it was with the beetle imitation that finally began to attract fish for me. In all likelihood it is a pattern which would not interest those who do not now have the beetle on their waters, though I have just heard this year that the beetle is now appearing in considerable quantities on the northern limestone streams, some one hundred miles away. So, for those of my good friends and all other good anglers who reside and fish there, the imitation that is herein presented may be of some benefit.

My favorite pattern was designed to represent the broad bulk of the natural, which is a rather thick insect. Any attempt to imitate this form with many layers of solid material such as wool yarn or thread makes it heavy and causes it to sink, time after time. Material such as cork, balsa, and the coffee bean are not easy to fasten securely, nor are they light enough to cast well. All of these were rejected in turn, and the bulky appearance was finally achieved by a more indirect method. It was done by using a large jungle cock nail, or a pair of them, tied flat over a hook shank thinly covered with black hackle, on the theory that a trout looking upward cannot see any dimension but the length and width of a flat-winged insect, never the thickness. The result is a large, flat, ovoid fly, very much like the natural and very light in weight. It casts like a No. 16 dry fly (which is

192

what it actually is) and floats beautifully, a solidly opaque creature.

For the underbody and for support on the water I have found nothing better than the greenish-black hackle from a Flemish Giant rooster, tied in open palmer fashion around the hook shank, then all of the hackle along the back cut away to make a flat table for the jungle cock nail, and a wide V cut away underneath to allow the fly to lie flush with the surface of the stream.

In the early development of this pattern I made the mistake of adding a body of peacock herl wound around the shank before applying the hackle, but I discovered that this was quite unnecessary and harmful, causing the fly to become sodden and difficult to float. The flat mat of short black hackle underneath the jungle cock nail makes a marvelously good underbody, from which the moisture is easily and quickly cracked to make it float. Do not think for a moment that I have deserted the principle enunciated in Chapter 3, i.e., that the bodies of terrestrials are of paramount importance. In this instance I have simply dispensed with unnecessary bulk and allowed the shapely jungle cock nail or nails to form an opaque background against which the screening hackle creates an impression of depth and bulk. In fact, the same idea is carried out in the case of the jassids, as the reader shall see. Finally, a most convincing addition was made in the form of legs, effected by the use of several strands of black ostrich herl fastened at the center of the hook shank and split apart by several turns of the ribbing hackle.

Now a word about the choice of jungle cock for this imitation. Much of this material is badly split and frayed, particularly the large nails; these may be perfectly satisfactory for salmon flies and such, but for our present purpose the less split the better. Perfect nails without splits are few, and one split will do no harm. The best-quality nails are usually found in the medium and small sizes, and these are to be preferred for the beetles. Two medium nails tied flat over the hackle and allowed to overlap slightly form a beautifully ovoid and durable body, and no wings are needed because in almost every case the living beetle falls to the water with wings tightly folded and invisible beneath the coppery-brown wing covers. If the reader is color-conscious, these wing covers can be closely imitated by choosing jungle cock with a decidedly orange cast, though this is only an approximation of the beetles' color; I will certify, however, that no more than an approximation is needed.

Chapter 8

GRASSHOPPER

I T IS with some diffidence that the grasshopper (Melanoplus Differentialis) is accorded a prominent position in formal dry-fly practice. The size of this creature and the attendant difficulty of using a comparable imitation on ordinary tackle ally its use more closely to the art of bass bugging than to that of dry-fly fishing. It must be admitted too that it lacks a great deal of the grace and refinement which accompanies the employment of dry flies in the 16, 18, and 20 sizes. Aesthetic values are rather low where the use of Melanoplus is concerned, and its ungracefulness is somewhat aggravated by its terrestrial origin and lineage. In all likelihood an imitation of this animal would not have agreed with the fine sensibilities of a man like Frederic Halford, who would

195

have complained, no doubt, that it was at variance with true dry-fly practice. His deprecation of so large an imitation as the Green Drake is indicative of his philosophy on this subject. I am in complete sympathy with his views, and would gladly trade the opportunity to fish to the grasshopper for that of fishing to the pale wateries, for example. Then, too, there is always a jarring note, a lack of harmony, associated with the ungainly efforts of even the most proficient caster in his attempts to make a smooth delivery of this cumbersome artificial. There must be many people, acutely aware of these differences, to whom the prospect of such fishing would be offensive, particularly those who delight in the oblique approach to the art of fly-fishing—the flashing elegance of the slender rod, the graceful curving movement of the line, and the fly falling like thistledown. None of this is probable with the artificial of Melanoplus; nevertheless, there are other considerations which must be taken into account, and they are such as might convert even a most elegant angler like Halford. There must be very few people who are immune to the powerful attraction of engaging with rod and line those behemoth trout approaching the five- and six-pound mark and even larger. The excitement of trading consequences with such trout would justify a slight lowering of the highest standards.

In those communities where the grasshopper exists in great quantities the dry-fly angler may anticipate the rarest kind of sport, similar to the incident described in Chapter 2, for these insects when present on the surface of the water constitute a toothsome persuasion to those sizable fish who do not normally concern themselves with the smaller insects. There are many parts of our streams which are flanked by close-cropped meadows in the lush farmland sectors—conditions which are suited to the presence and thrift of the grasshopper tribe.

196

In the early part of the summer they are not likely to be found on the water in any appreciable quantity. This is largely due to the fact that they have not yet reached the full physical development which permits them to make the long jumps or sustained flights which carry them to the water. In the undeveloped stage their activity is extremely limited, and only those individuals who are immediately adjacent to the stream sometimes commit an inadvertence by jumping the wrong way into the water. When this happens they are inclined to ride the water quietly, not nearly as actively as the mature specimens of midsummer. For this reason the trout, particularly the larger trout, are not yet aroused to that show of interest which occurs in late season. Here and there, at irregular and longish intervals, there may be an instance of rises to grasshopper but not enough to warrant the abandonment of a different pattern of fly. As the season advances, the grasshoppers' physical development and their increased activity become more and more apparent, providing the observant angler with a measure by which to gauge and anticipate oncoming events. Jumps become longer and flights more sustained, and more of the insects become water-borne by virtue of their incompetence in judging length and direction of jump or flight. It is a major tragedy of the grasshopper's existence that he must always jump or fly in the direction that he faces after landing from a previous jump, no matter what the cost, even at the expense of his life, expiring between the uncompromising jaws of a three-pounder.

When the full flower of the season has arrived, Melanoplus becomes a nervous creature, full of alarms and sudden impulses which cause him to leap away from the slightest disturbance. The blundering movements of cattle or the footsteps of an angler on the banks are oftentimes marked by a progressive rustling noise much like the sound of hail falling on cornstalks.

All of these are contributory factors which explain the presence of the grasshopper on the surface of the stream; banner times are in store for the fly-fisher on days when a strong wind is blowing across stream, coming or going, and the stronger the better. It may be one of those winds that drive a man to distraction if he is trying to cast right curve when the wind is blowing left curve, but he should console himself with the thought of the prospect of the wonderful fishing at hand, for the grasshoppers will be on the water in great numbers. Let Melanoplus make one of those sudden starts and he is hardly a foot or two off the ground before he is quickly steered downwind in a curving flight which more often than not deposits him willy-nilly on the surface of the stream.

It is interesting to note the behavior of these insects once they have alighted upon the water, and if this activity is carefully watched some clue may be gained which will assist the angler to employ the proper tactics in the presentation of the artificial.

In the usual case this insect seems to be immediately aware that he is placed in a precarious position, entirely out of his element, and no sooner does he make a landing than he begins to make for a safe harbor, vigorously sweeping the water on either side of his body with oar-like movements of the legs. Being a landlubber by birth and inclination, he does not understand the principle of lifting the oar at the end of the stroke but continues to thrash the water forward as well as backward, so that for every two inches of headway he gains he loses at least one. As may be suspected, his zeal is unavailing, and he gradually loses ground while the seduction of a powerful current draws him into full course, down a line of drift, and into the jaws of a hungry trout. On the other hand,

198

there are many times when he does not exhibit so much activity upon striking the water but lies quietly and floats away with little or no commotion at all. This is especially true in the morning, when the chill of the night air is still upon him, or when the day continues cool and there is no sunshine to warm him into a livelier state.

Both of these forms of conduct give rise to certain peculiarities which must be taken into account in order to make a proper delivery of the imitation, and they are not easily discovered elements in view of the fact that manipulation of the artificial is entirely different from that of the regular dry fly. Those anglers who adhere strictly to the principles of the dry fly must be prepared to abandon some of them when fishing to the grass-hopper, if they are to expect any measure of success.

In the first instance, when Melanoplus is very active on the water the trout are prone to display a rashness that is quite unlike their normal habits, and this appears to be more so with large trout than with small ones. No doubt the frenzied efforts of an escaping victim are exciting factors which cause him to forsake his customary caution and surrender to a stronger emotion without regard for consequences. His rises are no longer recognizable as such but become lunges and slashes to the right or left, forward or backward wherever the intended victim may be. Sometimes he may strike from a distance of four or five feet, then for change of pace he may suddenly revert to type and take one in a quiet and genteel way. This is a confusing state of affairs to the angler, for all of the principles connected with rise-forms, observation post, and taking position have gone agley. Ordinary methods are hopeless, rendering futile the lightest delivery, the dragless leader, and the longest floats, for there is a serious obstacle confronting the angler. It lies in the fact that the grasshopper has legs and they kick!

Therein lies the differential; no amount of jerking or hauling with the rod will simulate this movement.

The activity of the grasshopper on the water is characterized by short, jerky movements which cannot be simulated by an upstream cast because such a cast, as is ordinarily used in dry-fly fishing, causes a large part of the line to sag from the rod tip and this portion of the line absorbs the small motions of the tip without communicating them to the lure. To overcome this deficiency the tip must be moved in a long arc to tauten the connection; the result is a streaklike movement of the lure which seems to frighten and repel rather than attract. Plainly, the prime need is a method which would impart very small impulses to the imitation, making it float along with little fits and starts exactly the way the natural behaves.

It may not be amiss to recall again the account given in Chapter 2 of the encounter with the giant trout of the Letort. There is an additional item of information that two whole days were spent in trying to solve the problem posed by an artificial that does not kick. I was acutely conscious of this deficiency and labored incessantly to overcome it, making innumerable casts and trying every conceivable dodge to make the artificial perform in the manner of the natural. It was not until late afternoon of the second day I found the solution. It came about in this way: a very narrow line of drift was concentrated near the left bank looking upstream. A broad area of quiet, slow moving water lay toward the right bank. When the grasshoppers came down the line of drift, the ripply current caused them to bob and swing constantly in one direction or another while they desperately scratched the water in order to divert themselves to a safe haven. I reasoned that, if only I could take advantage of this ripply current to make the imitation perform

200

similarly, I would have a good chance of attracting the great trout. I persisted in my attempts to use this advantage but finally discovered, more by accident than design, that my error lay in the use of a slack leader. It happened that I made a very bad cast (I thought) resulting in a perfectly straight leader which was immediately seized by the fast current, causing a pulling and tugging on the imitation, making it swing to and fro as it bobbed down the ripply current. I had a bare moment of appreciation for this superior performance before a startling explosion occurred and I found myself fast to and joined in issue with a trout of unbelievable strength and size. The rest of the story has been told, but the lesson learned from this experience will bear repeating many times, for the execution in detail of that strange miscast of mine became the foundation and prescription for many successes thereafter.

Curiously enough, it is the first-rate caster who will have the greatest difficulty in trying to learn this method, for the long-ingrained habit of throwing a loose leader must be unlearned a little in order to cast the straight, taut leader recommended herein. It is another curious fact that accuracy in the cast is not a requisite as far as it concerns the position of the fish. Trout will come from surprising distances, from any direction, to take the artificial with long knifelike slashes on the surface of the water. No matter where the trout may be, the angler must take advantage of any concentration of ripply current which can be found within a reasonable range of the feeding trout, let us say anywhere up to five or six feet. Greater distances than this present a serious problem, for there are many stretches of water that are smooth-flowing and blameless of any suggestion of ripple or concentrated drift. In such cases it is quite obvious that the artificial, the leader, and the line together will

201

float at an even and similar pace totally unlike the natural. I ran into this problem many times and tried to overcome it by resorting to manipulations of the rod tip but without satisfactory results. Another accidental miscast supplied the answer that I sought.

A certain trout, a good two-pounder, occupied such an area as that described above. He was known to all of us as No. 51, having been previously caught, tagged, and returned for future observations. Since he was a notoriously free riser, many of us regarded him affectionately, for he was considered a likely candidate as a progenitor of a new strain of free-rising trout, a scheme which the Fly-Fishers' Club had contemplated for some time. For this reason everyone was encouraged to try for him as often as possible in order to discover the extent of his free-rising tendencies. It was not often that he could not be found feeding on the surface but he was nobody's fool, and it required a little more than ordinary artistry to cause his downfall. One day I happened to find him feeding on grasshoppers in this quiet stretch, and it was quite clear that he liked them kicking and alive. A dead float did not interest him. One of my casts fell on a spot where the direction of light enabled me to see clearly into the water. No sooner had my artificial landed than I perceived a golden flash darting upward from the bottom of the stream and my artificial was taken. The actions of the two principals, the landing artificial and the perpendicular rise of the trout were almost simultaneous. Another great lesson was learned, for it appeared that the only possibility of simulating the natural in these quiet stretches was the initial "spat" of the landing natural. But, note carefully, this must be done within close range of the trout so that he does not have time for the perusal that will resolve his doubt into a conviction of deceit!

202

In these smooth stretches of water the angler must literally try to hit the trout on the head with the imitation. It may be necessary, at times, to make many casts in order to locate a roving fish, but success with this method is surprisingly frequent. Another application of this principle is feasible where the grasshopper, because of chill or other indisposition, rides the water very quietly immediately after landing. At such times the trout show a more reserved attitude toward the natural and seemingly take them only after they have made certain that there is some evidence of life. In such cases a long float with the artificial should be avoided.

Much of the success enjoyed in grasshopper fishing depends on the construction of the imitation. It must be as light and airy as possible, yet preserve bulk so that the slightest irregularity in the current will impart a lifelike movement resembling the natural insect. The length and thickness of the imitation are important features, the former to insure a long axis, which heightens the swinging motion, and the latter to obtain greater displacement in the water, which makes for better floating properties. The quill foundation of the Bennett pattern is admirably suited to obtain those qualities.

The tactical problems discussed above would not be complete without reference to another situation which frequently arises and is connected with that circumstance where trout are rising in a backwater or quiet eddy. The cast which was prescribed for these places is not suitable when fishing to grasshopper. A slow, dragless float is not desirable, but the angler should purposely arrange his position so that he can cast across an intervening swift current, with taut leader so that the current will immediately impart a lifelike movement through the pulling leader. The dragging leader recommended in these cases is one of the few instances, to my knowledge, when trout are not adversely affected. Grasshoppers themselves often transgress lines of drift in their efforts to escape the water, and feeding trout seem to recognize this as normal behavior.

When trout are feeding close to a bank in the smooth stretches, it is a good plan to make the initial spat with the artificial between the trout and the nearest bank in order to create the impression that a live insect has fallen or jumped from bank to water. It is a most convincing piece of artfulness, for trout often occupy such a position at feeding time, as many know, and seemingly recognize it as an abundant source of food. In such places it is probable that desire and expectancy have already formed in the mind of the trout, creating in him a high pitch of preparedness to seize anything that drops, and his rise is likely to be sudden and explosive. It is a fine example of trout psychology upon which the knowledgeable angler must capitalize.

The extent and nature of fishing to grasshopper is not likely to be fully understood if comprehension is limited by the experience on the narrow confinements of the ordinary trout stream. A better idea of the grasshopper's importance may be gained by observations on a wide area of water such as the Susquehanna River. These pages are primarily concerned with the reactions of trout to the grasshopper, but the bass of the Susquehanna River also know how to appreciate this insect and it may not be amiss to render an account of their proclivities in this respect.

A few years ago, when my knowledge was less extensive on this subject, I was invited to join a bass-fishing expedition to the lower reaches of the Susquehanna. The excursion took place during the first week of September, a month after the trout season had closed and during the peak of the grasshopper's mating season. My prepara-

tions for this trip in the way of tackle included only the
conventional bait-casting rod and reel, with its comple-
ment of plug baits in various designs and actions. I had
no notions that anything different might be required, and
believed that my tackle was inclusive enough to meet
any disposition of the bass, be it top-water, mid-water,
or bottom. The other members of the party made earlier
provision for any eventuality by gathering live baits of
many forms which were traditionally acceptable and
enticing to the bass.

Upon arrival at the riverside we chose to fish an area
where a large part of the river, nearly half of its width,
converged upon a very narrow, swift channel of water,
pouring into a large, deep basin. I decided to fish along
the edge of the channel, thinking that it was a likely
place where bass might be waiting and ready to take
food.

Our arrival took place at noon and we fished for something like an hour without profit and without any apparent activity on the part of the bass. At the end of that
period a decided change became noticeable. Here and
there a flurry in the water attracted my attention with
startling effect, causing me to turn quickly in one direction or another as the source of the sound varied. The
flurries increased, then they became explosions, and the
explosions became a continuous roll as they multiplied
minute by minute. I was at a loss to account for this
astonishing event until one of our party suddenly shouted
that the grasshoppers were in flight.

I could not see clearly into the rough water of the
channel before me but finally began to see the insects on
the surface of the quiet backwaters and eddies. Then
someone pointed them out, high in the air and winging
their way across the river in every direction. With care-
full attention and some straining of the eyes it was pos-

205

sible to follow the aerial course of one of them until he could be seen to falter and finally plummet to the surface of the river. As noted before, this particular section of the river constituted a large area of water directed into a narrow, swift chute. Whatever grasshoppers fell into the broad area upriver were gathered together and funneled into the chute and the bass knowingly and with grim purpose arranged themselves along the length of this chute into a receiving line.

At the height of this wild carnival the bass seemed to be possessed by a madness that surpassed anything like their customary savagery. A number of times, when the grasshoppers were clustered in groups or four or five, I could see an individual bass explode upon one of these, then because of the rapidity of movement, he appeared to somersault in quick successive turns, taking a hapless victim with each turn.

If the bass were excited, it may be easily understood that we were too, and everyone busied himself with respective ideas and devices to try to make a record catch. But nothing availed us in this extremity, for we were totally unprepared for such happenings. The situation plainly called for a top-water lure resembling the grasshopper and fly-rod tackle to deliver it. Here, indeed was specific imitation with a vengeance. What avail then, stone catfish, minnows, hellgramites, and bass plugs?

I exhausted my selection of offerings and succeeded, by some strange chance, in taking two bass. No others were caught. One of the party opined that he had brought along a fly-rod outfit that might produce a better showing and we arranged this tackle as best we could in order to make a try. The rod was an extremely stiff and heavy affair more suited to contend with Atlantic salmon, and the line which accompanied it was an E level, resulting in a combination of such deplorable inefficiency that

206

it became a heartbreaking chore to make a decent cast. The only leader available was a four-foot strand of 4X gut and this was attached with some apprehension, for big bass are rarely taken with such tackle in rough water. A large fly was trimmed to resemble the grasshopper as closely as possible, attached to the gut and presented to the bass with as much skill as the tackle would allow. The inevitable happened. A large bass engulfed the pseudo-grasshopper, plunged into the heavy water shaking his head savagely, and quickly broke the leader without much ceremony. We were convinced of the futility of any more attempts of this kind and we concluded the day with a very light bag—two bass, to be exact.

On the following day I gave an account of this experience to an interested friend, and he undertook an immediate trip upriver with high hopes of enjoying some rare sport. I expected that his report would parallel mine in every detail but such was not the case. The grasshoppers were still there in great quantity but no bass were in evidence, and he spent a considerable amount of time rowing about the river in search of feeding bass, whereupon he made a strange discovery.

On that day, the clarity of the water was such that he could see bass resting motionless in places where the force of the current was negligible. The approach of the boat did not seem to alarm them in the least, not even when it was directly over them. Curious about this behavior or rather lack of behavior in a fish which ordinarily will not tolerate such things, he extended an oar blade and prodded them to see what would happen.  There was no response other than a slight shifting movement away from the blade. He concluded, and I believe correctly, that the bass had eaten so many grasshoppers as to create an excessive demand on their digestive powers, thereby reducing them to a state of stupefaction.

207

Autopsies on both trout and bass will often reveal the astonishing extent of their gluttony in grasshopper time. Their stomachs are gorged and distended so much that the wall tissues assume a semitransparent aspect, not smooth, but with a knobby surface caused by the outward pressure of the grasshopper forms.

Earlier mention has been made of the difficulties of imitating this insect and its importance to anglers during the latter part of the trout season. I wrestled with this problem for some time and made no real progress toward a solution until I took counsel with Bill Bennett of the Fly-Fishers' Club. Bill has achieved so much distinction in the art of fly dressing that I felt completely confident of the outcome when he accepted the gage and determined to perfect a workable pattern of the grasshopper. The imitation described herein is satisfactory in every respect and I would like to call attention to the pontoon feature, which in my estimation is a stroke of genius and a marvelous piece of engineering, insuring the likelihood that the "hopper" will float right-side-up and preserve body outline from the viewpoint of the trout. I felt that it was fitting and proper that Bill should write the tying instructions himself and herewith present them to the reader with my heartiest recommendations and my appreciation of the many exciting experiences the pontoon hopper has afforded me in the past.

THE PONTOON HOPPER

By Bill Bennett

THE dark secrets of stiff hackle, the fragility of weblike gut, the curling grace of a tapered line, and the visible rise of a power-packed trout to a high-floating dry will forever fascinate me. I'll admit that over the years I've been a worm or minnow dunker on occasion, an infrequent dabbler in the art of streamer and bucktail fishing, and a nibbler at the mysteries of the wet fly. It may very well be that my own lack of skill in these latter types of trouting has caused me gradually to abandon them and to concentrate on the dry fly. Don't misunderstand me; I'm not a dry-fly purist in the strictest sense of the word, I am just not interested in "slugging it out" with the wet stuff.

Feeling the way I do about the high-floating dry, it is easy to understand why I snorted in disgust when Charlie Fox dropped a 1½-inch imitation of the grasshopper in my hand last June, stating that he thought this thing had possibilities. If I recall correctly another gentleman who feels somewhat the same way I do about the dry fly also snorted and to show his absolute disinterest in the monstrosity proceeded to show us a dainty dry tied thorax-style to imitate the grasshopper. I'll admit that at that time I was a bit puzzled at Charlie's interest in this imitation of the "hopper," which in over-all proportions compared with a size 4 or 6 fly, since he, as a general rule, fishes them high and dry and then prefers 18's and 20's.

This outsize imitation, originally tied by Charlie Craighead, was promptly dismissed from my mind and would have remained thus had I not been subjected to almost daily association with Vince Marinaro and his tall tales of monstrous fish feeding continuously on live hoppers on the Letort and refusing all offers of well-fished dries. What I heard and later saw convinced me beyond a doubt that a properly made hopper could take these highly sophisticated Letort trout.

My interest in the original Craighead hopper had now

209

been renewed, and at this point I think it appropriate to explain further about the original construction of the so-called monstrosity and to make apologies to its creator, Charles Craighead, for any seemingly uncomplimentary, first-impression remarks; for, as it later developed, Craighead's original hopper with some refinement now has a definite place in my fly box.

The original hopper, as I recall it, was constructed of a goose or turkey quill about 1½ inches long tied securely on top of a size 14 or 16 hook. The quill was plugged at one end with cork or some other material, then wrapped closely with yellow tying silk and coated with clear lacquer. The wing, apparently a turkey wing feather, was folded back over the body and extended slightly over the end of the quill. Black hair, origin unknown, was used for feelers, and two brown hackles clipped close to the ribs and bent to simulate joints were used for the legs.

Craighead's imitation of the "hopper" proved effective to a certain extent. The Craighead imitation, however, had certain objectionable features listed as follows in the order of importance:

1. The all-silk wrappings and feather wings became watersoaked and caused the lure to sink.

2. The hopper had a tendency to float on its side, caused by lack of stability and aggravated by the wing placement, which was above center of balance.

3. The casting of the winged hopper was invariably accompanied by a whining and buzzing noise characteristic of large winged lures.

4. The clipped hackle used for legs presented a realistic appearance when freshly tied but straightened out on the first cast and trailed streamerlike thereafter.

210

My first attempt to improve Craighead's hopper resulted in partial failure, for my hopper, minus legs and plus oiled silk wings, floated on its side, whirled and buzzed disconcertingly while being cast, and was improved only to the extent that I had painted the quill body yellow instead of covering it with lacquered silk, thus eliminating the possibility of water-logging.

My next attempt was approached with a little more study. What we needed was a hopper that would float upright yet retain the general silhouette of the live hopper. The answer was the addition of two small quills on each side of the larger quill body—pontoons—which would serve a twofold purpose: first, to keep the hopper floating on even keel; and second, to simulate folded legs. A hastily constructed model compared with a live hopper quickly convinced me that my model closely resembled the general outlines of the real hopper afloat. That much behind me, I next turned to the construction of the body. To paint the quill seemed practical, but would it stand up on the underbody, which was to take quite a beating in long hours of casting? This problem, however, was solved quite accidentally; one night while rooting through my material cabinets I uncovered several dyed duck quill feathers in various colors. This was the answer, for I reasoned that if the quills were cut and prepared before dyeing the dye would penetrate both inside and outside the quill, thus assuring a lasting translucent body color.

I fully appreciate that the majority of our group do not tie flies, and that a description of the tying technique for the pontoon hopper, including the preparation of materials, might therefore prove uninteresting. For that reason, I have prepared the following descriptive sketches for use by our fly-tying members who might, some day, decide to tackle the tying of this hopper, which I have named the "Pontoon Hopper."

Tools needed, other than standard fly-tying equipment:

Jeweler's saw for cutting body quills
Long needle (crochet hook filed to a sharp hook)
Quill-cutting block (miniature miter box)

Materials needed:

Turkey quills for large hoppers—1½ inches and up
Goose quills for medium sized hoppers—¾ to 1¼ inch
Duck flight feathers (metallic-blue-tipped) for use in making legs (pontoons)
Testors quick-drying lacquer in the following colors: Green, Yellow, Brown, Black, White
Clear hard top coat lacquer
Moose mane for feelers
4X white tying silk
Wide gap hooks—sizes 12, 14, 16
Bullet-shaped corks—¼ × ⅝

After selecting the quills to be used, cut off the feather portion to within three inches of butt with a large pair of scissors. Cutting the quill to actual tying size with scissors is not recommended for the reason that the larger quills will split under compression, making them useless. After the quills are rough-cut, the next step is to cut them to tying lengths. This is accomplished by the use of a cutting or miter block which assures uniformity in length. After the quills are cut to tying length, the insides should be cleaned of all pith. This may be accomplished quickly by the use of a hooked wire or needle.

The body color of the natural grasshopper ranges from a bluish light green to greenish yellow. Variations of greens and yellows may be obtained with a little experimentation in dyeing time. I have experimented with various shades of yellow and green and find that the Tintex Maize Yellow and Nile Green dyes will produce the desired shades for use on the green- or yellow-bodied hopper.

Assuming that the tier has completed all the preliminary steps associated with the construction of the hopper, including the cutting and dyeing of quill, and that he has on hand sufficient bullet-shaped corks (which may be obtained from any fly-tying supplier), lacquers, and other materials hereinbefore listed, I now present the following step-by-step description of the trying technique for the hopper.

Cut off approximately ¼ inch from the cork and insert the cut, flat end of the remaining bullet-shaped portion in the open end of the quill about ⅛ of an inch. Since quill openings will vary in size, the ¼-inch cork may appear oversize. This is a distinct advantage, since the cork may be squeezed or compressed while being forced into the open end of the quill, thereby achieving a more secure fit. Corks may be cemented after compression, but this is not absolutely necessary. The final painting seals the cork and the quill permanently. After the quill is plugged, place it on top of a size 14 or 16 hook previously prepared with heavily waxed thread and wrap it closely.

The next step is to select two smaller quills which have been dyed a rust color for the legs or pontoons. These are compressed or flattened at the open ends and bound tightly to the sides of the body. Two moose-mane hairs complete the rough hopper; when they have been tied, cut off their stubs and finish off the thread at the eye of the hook.

Your hopper is now ready for painting. The first step is to paint the thorax, wing, and upper portions of the

MITRE BLOCK

SAW

TURKEY TAIL FEATHER

CORK ¼" x ⅝"

INSERT SMALL CORK AND GLUE

LASH ON HOOK

2 SMALL QUILLS

CUT

SMALL QUILLS TIED ON SIDES

MOOSE MANE

PAINT

213

legs olive green and brown. The hopper is completed with a few strokes of green and yellow lacquer at the head and by the addition of two black spots for eyes made with the heads of pins or brads.

Chapter 9

MINUTAE

IN ONE of the daily newspapers not very long ago there was an amusing cartoon of a fly-fisherman who succeeded in refining his terminal tackle so much that there was no longer any connection with the fly; only an aching void remained. The cartoonist must have known something about fishing with minute dry flies.

To see a trout rising to something invisible, to fasten a diminutive No. 22 dry fly to the gossamer point and cast that fly to the trout, judging the accuracy of the cast by following the line; to see the gentle swell of the rise again when the obscure No. 22 should have floated over the desired spot, then to tighten and discover the connection with a lunging trout is the most exotic experience that can befall a fly-fisherman. Let it happen a thousand

215

times or ten thousand times, the novelty of the event never palls, never loses that quality of breathless expectancy.

If the fly-fisherman were condemned to choose and be confined to a single phase of trout fishing for the rest of his life, and if he were limited to fishing for one-pounders, he could do no better than to determine upon this, the most fascinating form of fly-fishing in all the world. All forms of the dry fly have their peculiar attractions, but none of them begins to approach the ineffable charm of fishing with minutae, and nowhere is this sort of fishing more intensified or more prevalent than it is on the Letort and Big Spring at Newville. Neither of these two streams can boast of a decent hatch of large Ephemerids, nothing like those of the Yellow Breeches and the northern limestone waters, unless the transplanted Green Drake, now beginning to appear should supply this deficiency in the future. But for small insects in great quantity, terrestrial and water bred, they cannot be surpassed even if equaled elsewhere.

At Newville the best of this fishing occurs in the upper reaches, where the stream has a fine gravel bottom, the water is rather shallow, always gin clear, and the prevailing weed is watercress and chara. It is entirely occupied by Salvelinus, the native brook trout, and such trout I have never seen anywhere else. I never catch one of the better specimens but that it reminds me of Southcote's beautiful description, my favorite of them all, "its body of a fatness adorable, gleaming like copper above and like gold below."

Somewhere, somehow, Salvelinus has acquired a reputation for being a subaqueous feeder, not given to free-rising tendencies. This is a proposition that no Big Spring angler could ever understand, much less originate. Probably it grew out of earlier publicized experiences

216

Big Springs as it looked many years ago. The decorative
and historic old mill was demolished to make way for
a parking lot. Almost the entire volume of the stream
is represented by the spillway at the left and the water
issuing from the mill's archway just as it comes from
the caverns about one hundred yards above the mill.

Big Springs at McCollough's. Brook trout were alwa‚
very plentiful at the sides and below the small island
at the left and the larger island in the center. This
photograph depicts very well the rare and delicate beauty
of this lovely valley.

with the dark hemlock trout of the northern freestone streams and the earlier habit of fishing with wet fly only. Theodore Gordon knew better, a knowledge derived from his frequent visits to Big Spring and his broad familiarity with other limestone streams, as his letters would indicate. Everyone could have learned from him that Salvelinus knows how to rise for surface food and often, too.

I shall never forget the sight that I witnessed at Newville some years ago, before I knew what the minute insects meant in these waters. Accompanied by two companions, I spent the last day of the trout season at Big Spring. There were few anglers to be seen that day, strange to relate, and we were free to choose those sections that we liked best to fish. On that day, our preference settled upon the lowest reach of the shallow, gravelly area, sometimes called the divide, where the gravel bottom ends and muddy part begins. The valley hereabouts is low-lying and flat, affording long vistas unobstructed by trees and streamside foliage. Dividing the meadows between us, we separated widely apart in order not to interfere with each other in the event that it was necessary to range up and down the bank to find a feeding fish. Fishing was extremely poor all that day in spite of ideal weather and water conditions. Nothing moved. When evening arrived, a powerful wind came out of the west, driving across stream from the western bank with such terrible force that it picked up and carried with it all kinds of debris—leaves, branches, twigs, and the like. It was an uncomfortable wind that plastered the hatbrim against the forehead, filled the eyes with dust particles, and kept the fly line streaming and billowing behind the angler. It was hopeless to try to make a cast in the midst of this turmoil. Then the wind died as suddenly as it had begun; a period of dead calm set in lasting for the re-

217

mainder of the evening, and the water resumed its normal placid appearance. This change had barely taken place when the surface of the water became the scene of feeding activity the like of which one does not ordinarily encounter outside the bounds of a hatchery. As far as the eye could see, several hundred yards at least, the entire surface of the water was a mass of dimpling rise forms, occurring and recurring with increasing tempo as the evening advanced. Three or four hundred feeding trout was a fair estimate of those within my view. A like amount was estimated by each of my friends, making a rough total of one thousand or more brook trout rising at the same time but rising to something that was invisible to every one of us. That is no ordinary event. Trying as hard as possible, peering here and there in the half-light and prowling unceasingly up and down the bank in an anxious state of mind, I still could not determine the cause of this mysterious activity. I caught only one trout, a one-pounder, which I returned to the water. On comparing notes with the others at the end of the day, after the first tentative queries were warily exchanged disclosing a position that was equally embarrassing to all of us, I found that they had caught nothing and were as mystified as I was.

Not until I had seen it happen again on the Letort, not until I had seen and learned what is contained in the lush meadows of the limestone country—the little ants, the jassids, the tiny beetles, and all the other little beasties— not until I myself had caught trout in these circumstances and related these insects to those trout, did understanding finally come. Moreover, this understanding included a new meaning for the winds, nothing like what the ancient proverbs have to say. All winds are good winds, and like Father Izaak I shall henceforth "let the wind sit in what corner it will and do its worst, I heed it not."

218

It would not be strictly accurate to say that the importance of small insects was unknown to everyone. A few fly-fishermen, let us say, suspected their relationship to the mysterious rise-forms. There was one old fellow that I saw often on the Letort, a rather pleasant-looking but taciturn person, plying his long rod with the inborn grace and skill of a master craftsman. He was not inclined to extend himself in speechmaking; consequently it was a little surprising, one day to hear him come up behind me and make the remark that these trout "didn't like balloons." I was fishing a large bivisible at the time, something I no longer affect on these waters, and did not quite comprehend his meaning, but I could not mistake his aspersive tone. Somewhat nettled by his criticism, I asked him what he thought was better and he showed me a very small fly, perhaps a 20, a size that I believed unwarranted except for extra special conditions. I thought that he was eccentric and became convinced of this when a few days later I rounded a bend in the stream and found him fishing, of all things, a large Fanwing Royal Coachman, than which there is nothing *balloonier! !* I must have looked my surprise and my look must have prompted his rather apologetic explanation that his eyesight was so bad that he had to use something like that to see where he was fishing. The caustic remark that I was about to make died aborning and my sympathies were aroused, but I was still convinced of his eccentricity. I am sure now that he had many secrets to divulge about small flies, founded upon close observation and a lifelong association with these waters.

Both of the streams mentioned above can be properly fished with the same patterns of minutae, but there is a great difference in the method of approach, springing largely from the divergent habits of the trout and the character of the water in the respective streams. On the

219

Letort it is a rare necessity to wade the water in order to reach a feeding fish even against the far bank; although in a few places the stream broadens to a width of sixty feet, the general average is somewhere near thirty feet and very deep.

At Big Spring there are places as wide as two hundred feet, and it is sometimes necessary to wade through yards and yards of weed beds in order to reach the clear channels where the trout are lying and feeding. The labors of an angler ploughing through the weeds initiate a series of ripples or waves, following on one another and rolling over the feeding trout, giving them cause for alarm and bringing consternation to the angler. It is unavoidable; when finally a favorable position has been reached, there is nothing to do but wait patiently until their fears have quieted and their feeding has resumed. Even in the uppermost reaches of the gravelly section, where the stream is much narrower, the approach is equally hazardous. The stream being shallow, clear as crystal, and utterly calm in most places, a single footstep in the water, no matter how cautious, immediately starts the telltale ripples. On a few occasions I have tried the expedient of hurrying through the water with long, high steps, trying to keep abreast of the ripples while at the same time hurling a long cast to a feeding fish with the hopes of having him take the fly before the disturbance reaches him and puts him down. Believe it or not, this maneuver sometimes suceeds, but, sad to relate, there will be no more fishing in that area for a long time afterward. A better and far more effective approach can be made by pursuing a devious course through the weeds, placing each footstep behind a protruding clump of weeds so that the ripples are blocked and wasted before they reach the clear shallows. It is like the tactics of the Indian dissembling his person by dodging from tree trunk to tree trunk until he reaches a clearing and the object of his stalk.

Once having reached the desired position the angler is suddenly aware of the trout, plainly visible, swaying and balancing, naked to view in the most exposed part of the clear shallows. They do not seem to be alarmed but be not deceived. That short, leisurely movement upstream that they make is a sure sign that they see the angler. Nevertheless that is not discouragement, for these trout were not the object of the angler's approach. Let him reach forward to that further group upstream some sixty or seventy feet who are not aware of his presence. Let his leader be as long and fine as he can manage. Finally, let his artificial be one of the approved minutae patterns, and the chances of securing a rise are very good. Thus he should fish ride-and-tie, casting only to those alternate fish who have not been put down by his advances. It is the prettiest fishing imaginable. I know of only one other form of fishing that compares with it and that is the kind that exists on the commercial watercress farms at the source of the Letort.

Many years ago when the cress beds were open to public fishing I took every opportunity to visit them, and there I met, saw, and conquered the wildest fly-fishing of my career. The cress farms are composed of some acres of water impounded immediately below the exit from the underground caverns. They are really small, shallow ponds hidden by a dark green mass of thickly growing watercress. The whole area is divided into small square or rectangular plots bounded by wooden separators, distinguishing the sections in various stages of growth or harvest. The clear, sweet water continually seeps through the stalks of the cress, through the minor sluiceways, and into the major sluice which empties into the stream. The uppermost edge of the wooden sepa-

221

rators is very narrow, seldom more than a few inches, a bare catwalk, but sometimes enough to enable passage from one plot to another without any wading through the cress.

Imagine if you will, an angler entirely unencumbered except for net and rod, carefully picking his way along one of these catwalks, pausing now and then to insure his balance, then proceeding again until he stops with evident decision and adjusts his footing in preparation for some sort of violent purpose. What that purpose can be is not quite clear, for it is an unseemly way for an angler to behave and certainly no fit place for anyone to practice casting. But look a little closer into the dark green mass of the cress beds and here and there can be seen a sparkling bit of water no larger in area than a barrelhead. What mad purpose is this: the waving rod, the extending line, the fly dropping softly to the little pocket of water, and the angler gesturing frantically to maintain his balance. If his design is completed there comes a swelling and bulging in the little pocket, causing the adjacent cress to nod and tremble; the rod is bent, the line taut, and wonder of wonders! a trout comes thrashing and slithering across the tops of the thickly matted cress. That is "barrelhead" dry-fly fishing on the commercial cress beds.

All kinds of surprises are in store for the angler who attempts this kind of fishing. Many are the stirring battles that are lost and won, but mostly lost, particularly when the trout buries his head down and into the cress before he can be started across the top. Eventually the angler learns, by dint of many broken leaders and lost flies, that he can expect only a small measure of reward for his efforts. In time he discovers that no moment or fraction or a moment must be lost between the rise and the steady but tender pull which starts the trout across the

222

cress, keeping his head high until he enters the net without pause for reflection. Actually, no force is necessary, no more than enough to keep the trout's head up while the undulations of his body propel him forward.

I think now that the cress farms would be a wonderful place to try the smallest dry flies, since they are alive with the tiny terrestrials, the leaf-hoppers and such, as plentiful there as they are in the solid meadows. It is better, however, that the cress farms are no longer fished but are to serve as a splendid nursery and constant supply of small trout for the main stream. In addition to this, they are the last refuge for Salvelinus, the brook trout, largely superseded in the main stream by the comparatively new and satisfactory brown trout, Salmo Fario, the European. By and large this latter change has been more than less beneficial, especially in raising the average weight of trout which can be caught on minutae. One-pounders take consistently well, two-pounders less willingly, and a three-pounder is somewhat rare except in a place like the Paradise at Bellefonte, Pennsylvania, where incredible things can happen with fine gut and small dry flies.

Once, during the month of February when the snow was deep on the ground, I peeked cautiously over the parapet of the stone bridge at Bonnybrook above Carlisle and saw a fine trout, easily two pounds, hanging close to surface, just below a cloud of milling and dancing Diptera—the very smallest size, grayish in color, which thrive during the winter months but disappear when warm weather arrives. He was taking one after another with that distinctive rise-form, so trifling as to seem like the mark of a fingerling.

Normally, the rise to magnum food forms is attended by a more violent effort, causing a pronounced break in the surface, sending out concentric circles that expand

223

in all directions. The rise-form to minutae is defined as a triangular shape, the apex pointing upstream as though the trout had not used sufficient force to drive the concentric circles upstream as well as down. Immediately after the trout has taken, the triangular shape reforms behind the trout and becomes circular. It needs a keen and concentrated eyesight to detect this progression. Many of us only see the rise-form after the current has hurried it downstream behind the trout, and on that account we estimate the position of the trout incorrectly; this is the cause of casting short so many times.

Let the artificial be thrown a little further upstream than what seems to be justified, and allow a long float to circumvent the often repeated trait of Letort trout to delay the rise, bearing in mind, too, the adjustments to be made for the compound rise (the delay and lateral drift) or the complex rise (the delay, lateral drift, and reverse). A great many times, Letort trout like to maintain a halt in front of an obstruction, close to either bank, where the current is slowed, forming an eddy, and the drifting minutae are channeled into the trout's position. The rise is simple but leisurely made and the float of the artificial, though short, must be sustained by using the slack delivery.

For the patterns which give the best results, the imitations of the jassids and the ants (red and black), a small beetle form, and the little olive duns are sufficient. Some are inclined to question the worth of trying these patterns on hooks as small as No. 22, maintaining that an equally small fly can be tied on a 20 or 18, in order to gain a hooking and holding advantage. That is a good plan, as I can attest, but only for those insects which lie imprisoned in the surface film. In other cases they ride the surface so lightly that they do not create indentations, hence there is no light pattern and any excess weight in

A remarkably good photograph of the "V" or wedge-shaped riseform above and behind the trout immediately after he has taken a minute insect. Tiny bow waves in front of his nose and the spread fins indicate he is still moving back to his observation post.

An extreme close-up and a fine picture of the surface of the water at the exact moment when the trout has taken an insect, showing the depression caused by sucking and/or gravity. The shadowy outline of the trout is barely discernible. This is a part of the riseform that is almost impossible to see at normal fishing ranges. The camera reveals it very well.

The after-rise, showing the riseform hurrying behind
the trout, re-forming, and becoming circular.

The completed circle of the riseform, when it may be several or more feet behind the trout. The progression to these different stages is lightning fast. The great majority of fishermen see only the last stage, when the riseform is circular—and many feet behind the trout. You must be gifted with stop-action vision in order to see the early stages of the rise. Casting to the circular riseform is probably the gravest error committed by would-be dry-fly fishermen.

the imitation would render them distorted to the view of the trout. The jassid, in particular, needs special attention on this score.

The distinction between flies which ride flush or partly submerged and those which ride lightly is of such importance that success or failure hinges upon strict observation of this factor if the trout are accepted as final arbiters on this question, for they certainly appear to observe it themselves. There is no gainsaying the fact that these little hooks do not have desirable hooking qualities, but once they are embedded they hold quite as well as the larger hooks. Hooking qualities can be wonderfully improved by careful honing of the points as directed in the chapter on fly dressing. A point which is needle-sharp or "sticky" will begin to penetrate in the slightest contact and a moderate sneck or side twist is an additional advantage.

A satisfactory imitation of this sort is fairly easy to tie, but any experience with such a pattern will quickly reveal to the perceiving angler that a second important item decidedly affects the position on the water, namely, the weight of the leader point. The finest-size gut or nylon points normally available are not entirely suitable, for they cause the artificial to press severely on the surface film, partly submerging the fly and creating a questionable light pattern; 4X and 5X sizes are a pronounced disadvantage, while 6X gives fair results but leaves something to be desired where the lightest artificial is being used.

The problem of obtaining small hooks in the finest wire with good temper is a vexation at all times as many can verify. The search for the finest of gut in good quality is equally or more vexing. Once I had a stroke of good fortune in this respect, at a time when my constant queries about fine gut had become a tiresome refrain, even to me. At some point during the course of a luncheon at the Fly-Fishers' Club, the person sitting next to me, one Charles Knier, leaned in my direction and whispered something to command my attention. I did not hear him plainly, but could not doubt his manner for he wore a sly, Mona Lisa smile affected by those bearing delicious secrets; then slowly reaching his hand into his inner coat pocket and withdrawing it cautiously, he allowed me a momentary glimpse of the contents of a thin, flat packet. What I saw was enough to demand my entire attention. It looked to be silkworm, and such gut of the finest size I had ever seen. No more was said or done at the time, but then and there I laid swift plans to open negotiations for any or all of the precious article that I could acquire. After the luncheon, a meeting was arranged and effected and I had a chance to examine the gut, for such it turned out to be. It was genuine silkworm gut of the finest kind, 18 inches long and 8X or .004 inch in thickness, round and clean. I had never seen anything like it before. To my friend's everlasting credit he dealt with me more than fairly, and I came away the proud possessor of a dozen strands of this extraordinary stuff.

A trial with gut points of this size quickly demonstrated its superiority over any other size in executing a proper presentation of the dainty little dry flies. It provides an extra advantage—it eliminates much of the threat of drag since it falls more loosely than heavier and stiffer gut. Best of all, it allows the tiny artificial to float with a minimum disturbance of the surface film.

It is anticipated that many anglers will balk and question the practicability of such fine terminal tackle and, in truth, there is little encouragement to use it when one must expect break after break with fish of respectable proportions. It can hardly be denied that these are legitimate reasons for avoiding the use of ultrafine gut. A

226

diameter of .004 to .0045 inch permits an average breaking strength of approximately ¼ pound according to the N.A.A.C.C. table of standards—just four paltry ounces. The margin of safety allowed by these dimensions must be small indeed, particularly when it is sorely needed in those regular crises occurring at the time of striking and the first run of a hooked fish. Even a bellying line in a moderately fast current will easily absorb much of this margin and a break is always imminent. Nevertheless, a natural reluctance to discontinue the use of the fine gut and the strong desire to get the best results with the small flies encouraged me to continue the experiment. The proverbial cat has been skinned so often!

It would be plainly impossible and a ridiculous thing to prescribe a system advising the angler to employ a "delicate touch" and to gauge the resistance of the rod "within a hairbreath" of the breaking point of the leader. There are few people who could do that where a ¼ pound breaking limit is imposed. Many of us are heavy-handed and many of us become more so with the increasing fatigue of a long fishing day. For these reasons it was obvious that a formula involving definite and certain mechanics rather than vague, uncertain admonitions should be the proper foundation for the management of gossamer gut. That sounds as though it can be done with mathematical precision and so it can, believe it or not, but in a manner entirely unlike Father Izaak's quaint trick of throwing his rod into the water to float unhampered, when he feared a break with his horsehair leader, then following downstream and recovering the connection when the danger was past. That kindly patron of the angler's art never knew and never enjoyed the benefits of the modern reel with its fine adjustments for tension, though he once heard rumors that a kind of "winch" was used for salmon fishing.

227

But the reel alone, though efficient and capable of being regulated to the lightest uniform resistance, is not sufficient guarantee against the risk of breakage. There are other factors which nullify these benefits, and the suggestions offered hereafter to rectify them, though they seem to tread on the tender toes of orthodoxy, ought to be given a fair trial according to the directions which follow. Anyway, there is nothing sacred about tradition if it is a perpetuated mistake.

Let the angler make the experiment with any kind of fly tackle he desires; stiff rod or soft rod, light line or heavy line, even salmon tackle, it does not matter. No special equipment is required other than the refinement  of the leader. Procure a block of wood weighing between five and six ounces and a leader of sound gut or nylon tipped to 5X or .006 inch in diameter. Insert a screw eye in one end of the block. Reduce the tension of the reel to a one-ounce pull. With some reels it may be necessary to weaken the spring to accomplish this. Now assemble rod, reel, line, and leader and fasten the 5X leader point to the screw eye. Place the block of wood on a smooth, bare table top and retire to a distance of approximately 20 feet from the block, with rod in hand. Point the rod tip at the block, level with the table top, and reel the line taut. Now lift the point slowly and smoothly but *without touching or holding the line with either the right or left hand.* Continue to raise the rod point and eventually it will reach a position at 50 or 60 degrees where the line will cease to slide through the guides and the 6-ounce block will begin *to move across the table.*

How it is that a one-ounce resistance of the reel and free running line are able to move a six ounce block of wood? It is obvious that the resistance factors must have amounted to 6 or 7 ounces in order to overcome the inertia and weight of the block. Whence cometh the mys-

228

terious force to accomplish this? It comes, no more and no less, from the simple act of raising the rod tip, which causes the line to encounter more and more resistance to free passage when it is bent over a fixed obstacle. The more sharply the line is bent, the greater the resistance becomes, and the more it is capable of lifting. It is an old, established law of physics analogous to the principle demonstrated in the use of rope and pulley. What is most remarkable and alarming too, is this: that the act of raising the rod tip had wasted and absorbed almost the entire strength of the 5X gut, testing 8 ounces, leaving perhaps only a margin of safety of one ounce or less! It would be even worse if the rod point were lifted violently.

Now let the experiment be carried one step further. Lower the rod point to the original starting position, level with the table top and pointing at the block. Reel the line taut and without touching the line or reel handle with either hand, jerk the rod smartly backward (i.e., toward you), *keeping the rod point level with the table top*. The 5X point will not break! It is almost impossible to break sound 5X gut in this fashion. A great lesson is to be learned here, for it means that in actual fishing with very fine gut, the angler must, at all times, lower the rod to a horizontal position and point it at the hooked fish. What rank heresy! P. B. M. Allan, that self-styled heretic, could not have bettered it.

When this solution was submitted to angling friends in general, there was a storm of protest, and many arguments were advanced with vigor and sincerity against it. They resembled the traditional precepts of "keeping the rod tip high" to keep a tight line and to "cushion the shock" of unexpected lunges in one direction or another. These advantages are purely illusory even with the softest of rods, one that bends double from the weight of a

229

four-ounce trout. No matter how much it yields, it still provides greater resistance to free passage of the line than the reel alone with rod tip lowered.

This much is true and must be conceded: that the resistance to the running line created by the bent rod finally reaches a maximum that does not seriously affect gut in the heavier sizes, 1X or 2X for example. There is ample margin remaining for all practical purposes; but for gut sizes in 6X, 7X, and 8X it is extremely critical at all times. "Keep a tight line" is one of the ancient shibboleths that needs some examining, more pointedly for those who fish weedy waters than anyone else; its examination will result in a surprising discovery bearing upon the conduct of hooked fish.

The greatest of all hazards for the angler on the Letort or Big Spring is the inclination of trout, being hooked, to dive and burrow into the dense weed beds where they become absolutely immovable, even to the ultimate strength of the strongest practical gut. A break is inevitable unless the angler is willing to wade into the icy water and submerge himself to grub for the fish with his hands, a prospect which is by no means pleasant to contemplate, to say nothing of the risk to good health. There is no minimizing of the weed problem here, even for the fish, who have the greatest difficulty extricating themselves once they are enmeshed in the beds.

This past season a huge trout of some ten pounds was hooked in a clear channel at Big Spring. He tore about in the wildest manner, finally throwing himself out of the water in a long leap that landed him on top of an adjacent and thickly matted weed bed. He was trapped, and his violent exertions availed him naught for there was not enough free water to afford a purchase of his powerful tail and fins, all of them working furiously while his captor stared in stunned surprise at this un-

230

usual sight. The angler, finally overcoming his bewilderment, made a casual approach and easily netted the big fish.

Time after time I have witnessed similar incidents in which trout have been released, after capture, near a weed bed. If they are not replaced in a clear channel, they will remain on the surface of the weeds, exhausting themselves with futile efforts to attain the channels, and I have often been obliged to clear a path for them, urging and guiding them with my hands until they reached the clear water. Anglers of the northern limestone waters are not troubled nearly so much in these respects, those streams being, for the most part, swifter of current, and thus allowing less siltage and consequently less encouragement for weed growth than the southern streams. Imagine, then, the pleasure of the discovery that these trout, big or small, rarely or never dive into the weeds when they are managed in the fashion detailed for the use of fine gut. When the rod has been lifted gently after the rise to set the hook and then quickly dropped to the horizontal position, they seem to become confused, plainly uncertain of the source of the danger which threatens them, and they wander aimlessly about unable to resolve a course of action against this irritating but passive annoyance. Of course, they are not quickly brought to terms with these tactics, but an occasional and moderate prompting can be transmitted to the fish by carefully drawing the line taut with the left hand then releasing it quickly when the fish begins to move again.

The prime objective is keep him constantly moving, following him wherever he goes, never allowing him a moment of rest but never increasing his nervous pitch to the point where he becomes panicky and dives into the weed. Finally, when he exhibits signs of exhaustion by rolling thrusts upstream, exposing his underside, when

231

he begins to come to the surface and is no longer able to maintain his depth of position, then the angler may tempt fortune by carefully drawing the line tight with a delicate grasp of thumb and forefinger of the left hand, gently turning the trout's head toward the angler's bank but prepared to release the line at the first sign of panic. If he can be turned and started toward the angler, the real crisis is now at hand, for if he catches sight of the angler there is likely to be a tremendous surge of desperate activity ending with either a broken leader or the trout's entry of the weed. To allay his suspicions the angler should have crouched low before turning the trout's head and should remain perfectly immobile except for an imperceptible lifting of the rod to draw him over the submerged net. Any quick movement is fatal. Lastly, unless there is some urgent need to keep him as food, the angler should release him without hurt to the fat water whence he came, in gallant recognition of his sturdy resistance and his greater value in the water for the fascinating form of recreation that he provides.

If these directions seem unduly formalized and highly affected, let me validate them by referring to my fishing diary for a single season, the year 1945. During that year I made a total of 39 visits to the Letort, each of which I permanently recorded with respect to the number of rising trout that were observed, the type of rise-forms, and the class of insects being taken. A careful analysis of these records reveals the positive affirmations that on 20 of the 39 days or slightly better than 50% of the time, the trout were unquestionably taking the minute forms. These visits were not made on successive days but were strung out over the entire season. This was the most faithful account that was ever kept for a single stream, my having been occasioned by confinement to fishing the nearest accessible water when official restrictions were

Fox landing a fine trout for Ross Trimmer—seemingly from the ground. This site is part of the ancient mill-race that paralleled the main stream. It was thickly grown with watercress and other weed, but here and there could be found small areas of clear sparkling water where an accurately thrown dry fly would often tempt a fine trout. Note the small pocket of water in the lower left-hand corner.

And this is the trout.

imposed upon extensive travel during a critical shortage of transportation facilities; but there is no reason to believe that this was an exceptional year. As a matter of fact, other years, properly recorded, might yield an even more impressive tabulation in favor of minute insects, an assumption which is soundly based on the knowledge that the Letort, Big Spring, and a few other streams nearby are primarily small-fly waters, excepting only the presence of the sulphurs.

During the year 1948, I made an extensive effort to establish a more exact understanding of the small Ephemerids that inhabit these streams; and thereby hangs a tale of unusual interest, one that ought to capture the fancy of those who indulge in the entomological side of trout fishing. On April 12, just a few days before the opening of the trout season, a fine hatch of small mayflies was in progress on the Letort and I made a very good collection of this species, both male and female. This same hatch continued for several weeks. On May 15, a hatch of similar magnitude but a slightly smaller species was observed and another substantial collection was made. This hatch like the former continued for several weeks. On June 12 a third hatch took place and another collection was made. On August 8, a fourth hatch, as enduring and abundant as the first three, was noted and a fourth and last collection was made. In all four groups no spinners were available but the duns were allowed to molt in captivity, thereby providing the spinner, a more certain means of identification for the taxonomist.

Superficial inspection showed that all of them had a noticeable likeness in outward characteristics. They were uniformly slender, round bodied, and nicely streamlined, varying in the general color range of greenish olive to brownish olive; wings, ranging from pale smoky blue to a deep dark blue; legs and setae, a very light buff color,

233

sometimes almost white. Their very small size and subdued coloration made them quite troublesome to detect on the surface of the stream. Only by crouching low and moving about to catch the proper angle of light could I see them as small shadows against the mirror-smooth water, yet they were so plentiful that a hatch of Green Drake in equal numbers would have been cause for excited comment.

All of these things I had noted in previous years and I had often confided in Fox, Bennett, and others of my favorite company my belief that these particular Ephemerids were of uncommon significance in American trout stream entomology. My conviction that was founded on the conformation of the bodies, indicating perhaps a predun nymph of the swimming type, and the added fact that they seemed to repeat themselves over and over again during the entire year, as late as November.

After the collections had been properly prepared they were posted to Barnard D. Burks, associate taxonomist of the Illinois Natural History Division, Division of Insect Identification, who generously undertook the task of identifying them. Within a comparatively short time Burks's report was received; it divulged the astonishing news that all of these groups were species of the genus Baetis, species Vagans, Levitans, and Cingulatus. For many of us this terminology has little or no meaning, but a proper translation reveals that these groups are members of the same famous family that prevails in the English chalk streams, the olives and iron blues!

IRON BLUE

OLIVE DUN

As nearly as can be determined there has never been an American observer who has delivered more than a passing comment on the genus Baetis—a line or two perhaps, to note its scarcity, and an expression of regret that it does not abound in America, but no more. Yet here is a stream, in fact, several streams which have nothing but olives, practically speaking, the livelong year and in great quantity too!

Preston Jennings states that Baetis is "just about the scarcest fly to be found in America," that out of all the hundreds of flies collected on several streams, one single specimen of the olive dun genus (Baetis) was found. Wetzel does not catalogue even one of this genus in his *Entomology* and, more latterly, Flick is equally reticent.* Western authorities are no more enlightening and I have elsewhere stated similar views; but I am bound to say that the writers mentioned above are correct in their treatment of this genus, since it is purely a question of what Ephemerids are indigenous to a given locality, and the observer can report only what he has seen there. It was simply a matter of discovering their existence and confinement to these weedy waters, although the possibility that they are equally plentiful on other streams is not remote, with special reference to the northern limestone waters where they are most likely to be found.

The tally on this genus does not end with Dr. Burks's report, for this stimulating information encouraged some further research on the individual species and the outcome was the uncovering of Dr. Helen E. Murphy's remarkable study of Baetis Vagans contained in Dr. Needham's *Biology of the Mayflies.* Dr. Murphy reared two broods of Vagans in the laboratory and four successive broods under natural conditions in a stream near Ithaca, N. Y., and found that this species completes three life cycles in two years, thereby explaining the cause for the three periods of emergence of this species in May, August, and October of each season at Ithaca. I now regret very much that my collections ended on August 8, otherwise I might have found a similar circumstance on

* Jennings and Wetzel make some mention of Acentrella, a type of mayfly similar to Baetis.

235

the Letort, a fact which I had already suspected and which might have been reflected in Dr. Burks's report if only I had submitted specimens taken in the late summer and fall months. This procedure is indicated for the future; meanwhile I must suffer my impatience.

In addition to these findings Dr. Murphy established a positive relationship of length of nymphal life to temperature, but whether of water or air is not definitely stated in Needham's version of this study. I assume that water temperature was the important factor, since the nymphal period was the subject of experiment. The experimental data on this score shows that the Vagans nymph has a life cycle of six months when the average temperature is 60.7° F. to 62.2° F. and nine months when it is 41.4° F. to 47.9° F.

I have never found temperatures to be lower than 51° or 52° F. on the Letort where Vagans prevails, summer or winter, and never higher than 62° or 64° F. in summer; the temperature is held within this range by underground springs constantly supplying this stream. Moreover, the comparatively short length and great depth of this stream is not appreciably influenced by air temperatures. It has never been known to form top and anchor ice in the wintertime. It means, above all else, that Baetis Vagans of the Letort may enjoy a higher yearly average of water temperature, meaning further that the life cycle is accordingly shortened in the winter months, and warranting the assumption that there may be more than three emergence periods per season on this stream. This determination must await further investigation. The first collection of Vagans took place on April 12, and they repeated themselves on June 12, after the intervening hatch of Baetis Levitans, indicating a summer cycle of two months. I regret, as I said, that my collections did not extend beyond August 8, that it embraced a period of only four months and ended at the most critical time, when Vagans might have issued again. It was unfortunate too, that I made no collections during the month of July. It is interesting to note that Dr. Murphy's naturally reared broods began in May, whereas the first emergence on the Letort was a month earlier, and this may be offered to support the belief that the winter cycle of B. Vagans is considerably shortened hereabouts, as well as the summer cycles.

In any event, the answers to these questions may be of no more than academic interest to the general reader, since the late-season cycles may not coincide with the legal trout-fishing period in his jurisdiction. It is enough to know that Baetis exists, to provide a superlative kind of dry-fly fishing, and that they are a day-hatching insect, oftentimes beginning to emerge in the forenoon, but making their best show in midafternoon.

There is one fascinating habit of the little olives that must be mentioned. During the time of my collections and when I had the duns in the molting cages, I could not help but notice that the end segments of the bodies were constantly in motion, always twitching laterally, causing the setae to sweep gracefully and ceaselessly from side to side. In addition to this they are the best runners that I have ever seen in the Ephemeridae and never better than when they have become spinners, something that I had good reason to learn when I tried to handle them in the cages. It may be easily imagined how swift they must be as swimmers in the nymphal stage, probably as fast as any minnow. How discouraging it must be to a trout to try to make a meal of nymphs so small and baffling as these, particularly in weedy waters where they can dart about from shelter to shelter, with the blinding speed that they can effect.

The male spinners of all these species of Baetis are as

237

beautiful and dainty as anything can be, all of them having the clear, transparent middle segments, the deep brown thorax, and the rich maroon spot at the end of the body, much like the descriptions of the jenny spinner. The usual iridescence or play of colors—claret red, bottle green and blue—on the hyaline wings includes a peculiar bronzy gold glitter not ordinarily seen in other mayflies; but alas, an imitation of the male spinner is hardly justified, so seldom can they be found on the water, although I have worked out intriguing little patterns, just for practice, with the horsehair on bare hook as recommended by Halford. The female spinner of this genus is not nearly so handsome, generally being a dull brownish color in the body and slightly larger than the male, which is normal among all mayflies of any genus or species.

Fishing with spinners of either sex has been rather disappointing in the past, probably because they do not fall during the daylight hours, yet I have seen clouds of them at dusk engaged in the nuptial dance over the meadows. Perhaps they do not fall in the usual sense but crawl into the water to lay their eggs, as is customary with some mayflies. There is a very short paragraph in Needham's *Biology* which refers to oviposition by Baetis as it has been described by Eaton in 1888, Morgan in 1911, and Murphy in 1912. All of them agree that Baetis crawls into the water and deposits her eggs in a suitable place, preferably a stone, upon which she presses the openings of her oviducts; she then swings her abdomen from side to side, leaving the eggs clinging to the stone, while at the same time she crawls slowly forward leaving the eggs in a compact mass, made up of consecutive rows. What happens after that is a matter of conjecture, but it is possible that she returns to the air, flies about for a short time, then falls to the surface of the water.

238

This much I know, that they are often found drifting awash in the morning and sometimes afford an excuse for the earliest rises in the forenoon. I have seen heavy clusters of the spent females clinging to some old pilings that were rotted and decayed, a onetime foundation for an ancient bridge. Whether or not they used the pilings as a depository below the surface I did not determine. Halford states positively in *Modern Development of the Dry Fly*, Chapter IV, that some individuals of a species deposit their eggs by hovering over the water and dipping down at intervals to wash off a few eggs at a time, and that others of the same species will set their wings back, bring them together at the tips to enclose a small air bubble, crawl down a post or blade of grass or reed into the water, and deposit their eggs at the spot they have selected as being most suitable. I had supposed that the egg-laying habits of any species were invariable, but I can offer nothing to contravene Halford's assertion without making further inquiry along these lines.

I fear that everything submitted here concerning the Letort olives is too meager and nebulous to satisfy the reader; nevertheless, the problems have been stated and they mark the outlines for future study, which should include a determination of the ecological factors most suited for the olives' increase and perpetuation. The olive is just another of the many charming aspects of the Letort; and this discussion of it may help to afford the reader a better understanding of why we have deluged him with so many references to this unique little stream, and perhaps, too, a pang of regret that half of its length, the best half, remains untenable for the trout, having been violated by unsavory discharges from various sources, although there exists a feasibility of restoring it again to its original purity.

In spite of all these uncertainties about the spinners

239

it will do no harm to carry a few patterns of the imitation, and to try them early in the day or when other small spinners of a different genus are on the water and trout are rising to them. In the last days of the 1947 season I witnessed a circumstance when the latter condition prevailed, on Penn's Creek just below the outlet of Cherry Run near Weikert. Those who are familiar with this area ought to recall the flat, smooth character of the water and the great width of the stream in that neighborhood. It has been a famous piece of water for many years and a favorite resort for many fly-fishermen during the annual hatch of the Green Drake.

With Fox and Craighead for good company, I wandered up and down this stretch, looking closely for any sign of trout or hatching Ephemerids and trying to learn more about this interesting stream, with which I have never been too well acquainted. None of us had any luck with the fish that afternoon; then along toward evening I met an inhabitant of that locality who willingly offered some pertinent information on the fishing outlook. He complained, somewhat dejectedly, that the trout had been rising like mad in the several evenings preceding this one, yet no one had been able to take them, explaining further that the trout positively ignored the large mayflies of the instant season and were taking something that no one could see. He added that it would have done no good to see them anyway, for no one in that district used anything smaller than a size 14 in dry flies. I was far more heartened by this news than my informant would have believed, especially because I relish nothing better than to fish with the minute stuff.

There were two fine duns hatching at the time, a very large one that looked suspiciously like an Isonychia and a smaller dun with the bluest wings and the greenest body I have ever seen. I would certainly like to know

240

more about the latter insect. I disregarded both of these for the time while I bent my attentions to find some sign of small insects, finally discovering what seemed portentous—a tiny spinner, glassy-winged and brown-bodied, of lineage unknown to me. While pattering about in this fashion I heard a commotion behind me and turned to find Fox a short distance below me, in midstream, seriously contemplating the antics of three trout feeding in mysterious style on nothing that could be seen. I sat down on a boulder near the bank to watch the show, since Fox is a very good performer and I was curious to see the outcome.

In due time he made several casts without reward, then a bit of study and a change of fly; then more casts, more refusals, another change, and finally success in raising and hooking Trouts 1, 2, and 3, in faultless order. It was an entertaining performance, the more so because he had employed the process of diminution, a favorite trick of his; that he had done so was confirmed when he waded to shore, approaching me with a knowing grin, and holding up for my inspection a tiny No. 22 dry fly. Minute insects are no mystery to him. I had no opportunity to emulate him, for in the meantime the thunderheads had been piling over the mountains, accompanied by ominous rumblings, and before long we were hurrying to the shelter of the automobile, which we did not reach without the discomfort of a severe drenching.

There are times when a trout can be inveigled with a small pattern of generic design, other times when an even larger fly might interest them, but never, never be deluded into such a position when they are taking only the ants. Dogmatic assertions of this sort are dangerous things to use in fishing talk, but this is one that I cannot honestly avoid, having been forced to express this opinion by the memory of numerous defeats when I did not

241

use an adequate imitation. Elsewhere the theory and directions for its dressing have been given, though I would be the last to say that they cannot be improved upon, particularly on the score of materials. I have used alternative dressings, but I have always maintained as faithfully as possible the form and light pattern required for this insect. The earliest successes with the pattern were tied with black bear's hair, and a good pattern it is too. The long shiny guard hairs were cut close to the skin in a small clump, then twisted slightly to form a yarn and wound on the hook shank as though it were floss silk, to shape the large and small knobs. Its great fault is a tendency for the turns of hair to become loose and fall backward over the bend of the hook.

Others were tied with black spun fur and gave equally good results, although the fur lacks the hard shiny appearance that can be obtained with natural black horsehair, my favorite material at the present time, which is very durable and floats consistently well in the small sizes, if not quite as well as the first two. It needs to be dried carefully and delivered softly so that the spent wings do not break through the surface film, thereby losing support. To date, it is the best of all the patterns that I have tried, despite its drawbacks. Fox prefers the body of spun fur, probably because he has made some record catches (a good reason) with this pattern, the best so far having been made this past season on Oliver Deibler's water on Spring Creek, the big northern limestone stream and one of the finest in the land. We had planned to make the trip together, pursuant to Mr. Deibler's generous invitation which no one should ever refuse, for his stretch provides the finest kind of fly-fishing over big fish that can be found. I was especially anxious to try this fishing in order to make an exhaustive trial with the ultra-fine 8X gut. At the last moment, before departure, a

242

change of plans was forced upon me, no small irritation, and I had to forego the visit. Fox made the journey alone, returned in due time, and reported an experience that aroused my uncontrollable envy, namely, that he had raised, hooked, and then landed some 35 or 40 trout in a single afternoon, among them some three-pounders, all of which were returned to the water except a few in accordance with Mr. Deibler's wishes, and all of which were taken on the 8X gut and the black ant tied with spun fur. As I recall, he had a few unavoidable breaks with large fish that had reached the shelter of some underwater hazards, but the general results with the gut and the ant were eminently satisfactory.

These same considerations apply to the red ant except that the body must be as translucent as possible; for this purpose I have found nothing to equal the fine horsehair supplied for violin bows, washed and cleaned over and over again until it is crystal clear, then dyed a pale golden-brown color, to look like sparkling globes of light-colored honey, when the double-form body is constructed.

The last of the most worthy patterns of the minutae are the jassids, a name that has been dropped and replaced by the term Cicadellidae, recently revived by the entomologists. Though this name has a liquid and lilting sound, it is an effeminate and maudlin thing that I cannot stomach no matter how much the revision is justified. So contrary to the wishes of the entomologists I shall continue to use the former name, a compact and more vigorous term befitting the nomenclature of things connected with outdoor activity—things like rod, reel, and trout.

JASSID

One of the more troublesome problems connected with the smallest of imitations, particularly this one, is the matter of eliminating the light pattern, something which

243

is incompatible with the appearance of these tiny creatures on the surface film. Hooks of the finest wire and good temper are an invaluable aid, thinly built bodies are another, and wings tied flat over the thin bodies supply the requisite impression of bulk since trout cannot see things in three-dimensional terms anyway unless they are gifted with the power of imagination. In addition to these aids, I have suggested the use of hackle to obtain the maximum support with the fewest of fibres, accomplished by tying in at the bend of the hook and turning in the manner of a ribbing hackle, making one complete turn at the bend, a half turn at the middle of the body, and one complete turn of the head, or 2½ turns in all. It is an effective method and helps to make them ride very lightly with a minimum of disturbance to the surface film.

Those who are fortunate enough to possess a good copy of Ronald's *Entomology* (it need not be an early edition for some of the later ones are beautifully done) will find, I am quite sure, a very good illustration of this insect under the name "Wren tail" (frog-hopper, bent hopper, and the like); and Ronald, too, takes note in his rather sketchy comment that there are many species and colors of this insect, though he prescribes a definite color scheme which can be no more than the mildest of gestures to thwart the futility of standardization. He notes, too, that they are "very busy on hot days, hopping about and taking flights of about twenty yards and this is the time to use the imitation, for they sometimes drop short and fall upon the water. In colder weather they are found upon the long grass principally, not much on the water." It is a terse and meaningful paragraph containing almost the whole philosophy of fishing with terrestrials, and I would add the extra important constituent that they provide the best fishing when they are sub-

244

jected to the fiercest, most violent winds bearing them as largesse to the hungry trout.

And now, they are completed, to wit, my "jury of flies fit to condemn every fish in the river."

Chapter 10

FLY-DRESSING PRACTICE, or Imitation
Gained

T HE tools required for our purpose should be the best
that can be obtained. There is a wide variety and choice
of equipment on the market these days and most of it
seems to be of good quality. Vises should have small neat
jaws that will not interfere with the operator's fingers or
his vision, and in spite of much opinion to the contrary,
I like them with soft steel jaws rather than the hard,
tempered ones. The latter type require much more pressure to keep the hook from slipping down or up because
of the hard unyielding steel. The softer type become
slightly scored and dented from use and can hold a high-
tempered hook lightly without breaking it. A jeweler's-
pin vise served me well for years; I never saw anything
which held small hooks better, largely because of the

247

soft steel jaws. But such a vise does not compare with the excellence of modern vises in design or efficiency.

The most important tool of all in my estimation is the hackle pliers. These should have small, narrow jaws, hand fitted. Do not hesitate to purchase the finest that can be obtained. Nothing is more exasperating than to come to the last turn of hackle on your finished fly and have the hackle point slip away from the jaws. For the rest, a good pair of small scissors and a pair of surgeon's curved forceps are all that I require. I am aware that there are many other devices which have been invented and perfected to aid the fly dresser, and no doubt all of them are useful. Let the amateur or professional suit his individual taste in the matter; and I am bound to say that he is justified in using anything else that will help him to turn out a neater, better fly. In my own case I find one other item of equipment very useful at times. When I am using fine silk, such as 6/0, I often experience breakage, particularly when I am out of practice; it takes some time to learn to gauge the correct tension. In order to overcome this difficulty the flytier may use a bobbin, allowing the weight of the bobbin to supply the necessary tension, just short of the breaking point. In this way uniformity of tension is obtained and the silk is rarely broken.

HOOKS

Everyone who has dressed and used dry flies has suffered the annoyances that are often encountered in hooks of poor design, poor temper, and bad workmanship. The search for good hooks of light wire, good temper, fine points, and closed eyes never ends for any of us, and when a batch of fine quality is found it is cause for much rejoicing.

248

There is one feature about all hooks that I have discovered and do not like. None of them has a point sharp enough to suit me. In much of our limestone fishing small flies on small hooks (18's, 20's, 22's) are the rule, and a small hook does not have good hooking qualities, although once it takes hold it will dig deep and hold as well as a large one. In order to improve their hooking qualities I have taken to honing the points on my hooks before using them for fly dressing. For this purpose I place the hook in a jeweler's pin vise with the eye to the right and rub the point with a pencil-point, hard Arkansas stone, held in the right hand as though one were holding a pen or pencil. Rub the point from every quarter, top and bottom and the sides, always bearing away from the barb, turning the vise with the left hand when it is necessary to expose a new surface. Then the following test is applied. Place the point on the thumb nail very lightly and draw it toward the edge of the nail. If the point feels "sticky" and bites into the nail it is sharp enough. If it slides it is necessary to hone some more. After examining thousands of hooks I must say that I have never seen a single one that would bite into the nail without being honed.

The professional and some amateurs are likely to balk at the idea of going to all this trouble, since it is time-consuming and tedious. This may be so, but if small hooks are used this procedure will pay enormous dividends. A hook which is honed in this fashion will search and penetrate any part of a fish's mouth from the sole pressure of the jaws without any necessity of striking. If you squeeze such a hook, on the flat sides, with the forefinger and thumb the point will catch and stick in the skin without any further assistance, and this is a tremendous advantage when using small hooks.

The design of the hook used in all of the patterns

should be one with a downeye and a squarish bend, one such as the Model Perfect. It goes without saying that light wire is desirable in dry-fly hooks; there need be no fear of using it, if its temper is good, even where fairly large trout are concerned, although I find that coarser wire in small hooks does not affect their floating qualities appreciably.

TO TIE AN UPRIGHT WINGED DUN, THORAX STYLE

The instructions which follow are written from the viewpoint of a right-handed person. Left-handers can reverse the procedure and adjust it to suit themselves.

THORAX

TYING SILK

Place the hook in the vise with the eye pointing to the right. Wax about two inches of the tying silk or use varnish if you like. Start the tying silk in the *center of the shank* by doubling it back on itself and making 3 or 4 turns, allowing the silk to hang taut from the weight of the bobbin or hackle pliers. If there is any doubt about the location of the center of the shank, start the silk a little closer to the bend than the eye. This is important.

The hackle-point wings may be mounted together or separately, by holding the tip of each wing with the thumb and forefinger of the left hand and taking two turns of the tying silk around the stem of each in turn. As a matter of fact the two turns around the stem of the second wing will amount to four turns on the first wing. Bend the stems back toward the bend of the hook and make two or three turns behind the wings. Clip the protruding stems close and continue the tying silk to within 3 or 4 turns of the point where the bend of the hook begins to curve.

WINGS CUT FROM HACKLE POINTS

Choose 4 fibres of spade or throat hackle of the longest, glossiest, and stiffest nature possible; lay them on top

250

of the hook and anchor them with 3 or 4 turns of the silk. They should be tied in at the extreme butt end in order to obtain maximum length of stylets. In case the butt ends have a little web, tie down where the webby part ends and the glossy part begins. Now bring the tying silk behind and underneath the tail fibres 3 or 4 times, pulling each turn toward the eye, thereby lifting and spreading the tail fibres above the level of the shank, then split them widely apart in two pairs with 2 or 3 figure-eight turns between them and a final turn underneath to keep them high.

Next, take a piece of spun fur 6 or 7 inches long, using only one ply; it is usually 2- or 3-ply, but the plies can easily be separated by pulling them apart slowly and carefully unwinding them a little as they are pulled. Lay one end of the fur on top of the hook shank, allowing it to extend along the shank as far as the wings. Start the tying silk as close to the tail fibres as possible and wind over the fur up to the wings and one turn in front of the wings. Let the silk hang there. Now take the fur and give it a few twists in the direction in which it was originally twisted. This will prevent it from coming apart during the winding. Continue to wind along the shank, twisting as you go and laying one turn in front of another up to the wings, then several extra turns behind and in front of the wings. These last few turns at the base of the wings are important, since they fill the gap underneath the wings and provide anchoring shoulders for the hackle. Hold the last turn of the fur with two or three turns of the tying silk in front of the wings.

TIE IN SPU
FUR AND
WIND FRO

Now take two small hackles, one of them a little shorter in the fibres than the other. Anchor the stubs in front of the wings with 3 turns of the silk, with tips pointing toward the bend. Take the long-fibred hackle and make two turns, winding them in front of the wings on

TIE IN
HACKL

251

HACKLE
DIRECTION

top of the shank and behind the wings underneath the shank, thereby setting the hackle at an angle of about 45° to the shank. Tie in the hackle tip with 2 turns of the silk as close to the front of the wings as possible. Now take the short-fibred hackle and reverse the angle of the first by winding 2 turns on top of the shank behind the wings and underneath in front of the wings. In this fashion the two hackles are crossed in an X design. Tie in the tip in front of the wing with the turns of the silk as close to the wing as possible. Turn the tying silk up to the eye and allow it to hang.

Take the fur, which has been allowed to hang free and should not have been cut, and start to wind in front of the wings and hackle toward the eye but one turn short of the eye. Wind back again toward the wing as close as possible, then forward to the eye, and tie off at the neck with whip finish or half-hitches, whichever is preferred.

WHIP FINISH

If one is fortunate enough to possess long hackles with short fibres, one of these may be sufficient instead of the two recommended. In such a case, make the first few turns in front and on top of the shank and behind the wings underneath the shank, then reverse the direction of the turns by winding on top of the shank behind the wings and underneath the shank in front of the wings. This procedure will allow the long fibres at the base of the hackle to point toward the bend and the shorter fibres near the tip to point toward the eye. It is absolutely necessary to use shorter fibres in the fore part of the fly to prevent the fibres from bending backward sharply as a result of repeated pick-ups from the surface of the water; more important still, it will help to make the fly tilt forward a little when it is cast on the water, thereby insuring the likelihood that the tail will be elevated. In fact it is a good plan to clip them a little in the fore part if short-fibred hackles are not available.

252

In building up the thorax portion of the body take care to make a smoothly tapering outline, rather thick near the wing. Repeated references to the drawings or a natural insect will eventually school the fly dresser in the correct proportions which are desirable. Do not attempt to obtain a segmented effect in the thorax, since this part of the natural is rather smooth and unmarked. It is best to handle the fur with very little twist at this point, lapping the turns as flat as possible and using only enough twist to keep the fur from pulling apart.

In the case of the minute duns, notably the olives, the technique is exactly the same as for the large duns except that the hackle, one only, is tied in with the spun fur at the bend of the hook and turned the required 2½ turns as though it were a ribbing hackle, but turned after the body and thorax are completed. One complete turn should be made at the tail, a half turn at the middle, and one full turn at the head. Thus a minimum of hackle is used to the best advantage without overdressing these very small flies.

WINGS

The evidence previously offered to support the great need for a proper wing in the dressing of duns is overwhelming and should not be taken lightly. The great *desideratum* in fashioning these wings is to obtain the clean, flat appearance as it occurs in the natural fly. I have recommended the use of cut and shaped hackle wings, taken near to the top of the webby part, and I know of no better wing, but the fly dresser is not limited to hackle alone for this type of wing. They can be cut from the tips of duck breast feathers or the small covert feathers at the base of a duck wing. In every case they should be cut out of whole feathers in order to obtain the supporting center rib. They are remarkably durable and effective. A fine wing—for example, on the Cahill type of fly—can be shaped from two whole lemon wood duck feathers, though this is an extravagance against which many would rebel. In all of these cases they can be cut and shaped with a fine pair of scissors, but for the sake of uniformity and speed a cutting tool can be made which fashions a pair of wings with remarkable ease. A wing cutter of this nature was designed and perfected by two members of the Fly-Fishers' Club and it has been very successful and pleasing to use.

TAILS

Fly dressers have never really profited from the enormous mechanical advantages to be gained by a proper arrangement of tail fibres. Four fibres only, spread widely and maintained in that position, will support a size 14 to 20 fly better than a dozen fibres bunched together. The bunched fibres are all wrong, not only in practice, but in principle also. They cause capillarity, which makes them sodden and difficult to dry. Two or four fibres, split widely apart and fixed in that position, act like the outriggers of a canoe and guarantee that the fly will alight correctly.

In the case of the thorax style of dressing the tail fibres seldom touch the water, but in the event of a faulty cast or a badly balanced fly the split tails are always there to act as governors; and even if they touch the surface film, there is no appreciable alteration of the light pattern. This arrangement of the tails is undoubtedly the greatest single factor which insures proper positioning of the fly on the water. Successful cocking of the fly is almost a sure thing on every cast, even in a high wind. The fly dresser should practice this detail until he has mastered the technique thoroughly, and he should keep on hand at least one cock's neck of neutral shade (honey dun or blue dun) and of the best quality, with long stiff fibres to be used for tails only. Only one color of hackle is needed for this purpose, but almost any color is useful if it is available in good quality.

One of the best methods of fixing the tails in the forked position is to use dry silk, then after the fibres are split and secured with figure-eight and supporting turns, to place a tiny drop of varnish at the root of the tails. It will harden and shrink the silk, thus making the whole arrangement a permanent fixture.

Of course, as I have hereinbefore stated, the relationship of hackle and tails is not the same as it is in the conventional dressing. Tails are not a main support in the thorax style of dressing, but it is anticipated that sometimes the fly dresser will fail to angle the turns of hackle properly and the artificial will not sit up without the support of the tails. There need be no great concern about this matter, simply because the hackle is placed so close to the center of balance that the lightest support of the tail fibres will keep the fly in the correct position.

BODIES

Once again I would like to repeat and emphasize my conviction that the imitation of bodies on the *dun or subimago* of the mayfly is meaningless and superfluous. I have no hesitancy about taking a strong position on this question even though there are statements to the contrary by respected authorities. I know for example that some like to make a distinction between the male and female of the Hendrickson dun and like to tie these to differ, in body color especially. There is no good reason for such a practice and there are a lot of reasons against it. This criticism is not offered to disparage or discredit, but to suggest that perhaps others are honestly mistaken in their appraisal of this matter.

Many instances are cited to prove that a trout would accept an artificial dun with a particular body color after many other flies failed. This may seem like a justification for such a conclusion, but there are other decisive factors involved in these instances which seem to be disregarded. Sometimes it happens this way: a sodden dry fly, dressed with a heavy wool body, is exchanged for one dressed with peacock quill, a substance which is waterproof and extremely light in weight. Wool-bodied flies are notoriously poor floaters, quite unlike fur in this respect, and the result is a dun pattern which floats awash, thus completely distorting and differing from the light pattern formed by the surrounding naturals. Upon being presented with this quill-bodied pattern the trout rises willingly, whereupon the fisherman congratulates himself and concludes that the body was responsible for the difference in results. He is right, too, but not for the reason that he thinks. What actually happens in such cases is that there is a more perfect rendering of the light pattern by the quill-bodied fly because, being lighter in weight and almost completely waterproof, it allows the body to remain raised above the surface of the water, not touching, thereby permitting only the points of the tail fibres and hackle to disturb the mirror. This is all that is needed to induce a trout to rise, and if the wing be adequate when it appears in his window, he will take the fly. The truth of the matter is that the trout was not influenced by the different color of the body but by the fact that there was no body at all, since it is invisible outside the opaque mirror. When he rises to take, in this window, it is too late for him to quibble about bodies. In addition to this and all that has been said in Chapter 3, there is

256

a wealth of overwhelming evidence against the use of bodies in the imitation of the duns.

If bodies meant anything at all, Ed Hewitt would hardly ever take a trout on his famous Neversink Skater, which imitates nothing but the light pattern and a wing— a perfect fly for dun fishing, and one which limestone fishermen acclaim as one of the finest, if not the best, for Green Drake fishing. But alas, it is probably the most fragile of all artificials and is almost completely ruined after taking a fish. In fact, it does not withstand very much casting. In order to use it successfully, the fisherman must carry it by the boxful for frequent replacement. Neversink Skaters are the most exasperating in evening fishing, especially in June when the madness of Green Drake time grips trout and angler alike, rendering the latter almost devoid of self-control and ability to tie useful knots during the heat and press of the awful excitement. But if you must have bodies on the imitations of duns, and if it offends your taste (as it offended the taste of Halford and Marryat) to dress flies without bodies, then choose the lightest and thinnest of materials to allow your fly a chance to float properly, i.e., to manifest its presence to the trout by a light pattern only. In the descriptions of the duns included herein I have included a body material and appropriate instructions for its use. Although it could just as well be omitted, nevertheless, for those who insist on using it, the spun fur and its management recommended herein will make a light and serviceable body, but it should be applied as thinly and sparingly as possible. It is my practice to draw it between forefinger and thumbnail repeatedly in order to make it thin, hardly larger than the tying silk itself, and I like to make several extra turns, figure-eight fashion, at the base of the wings in order to make an anchoring shoulder or thorax to hold the hackle at the proper

257

angle, a purely utilitarian purpose. The body of your dun in front of and behind the thorax should be no more than a very thin bit of fluff, rather misty-looking and indefinable.

HENDRICKSON DUN

Wing: Darkest slate blue—almost black—cut and shaped from the webby part of 2 large neck hackles.
Body and Thorax: Reddish-tan or dark yellow spun fur.
Hackle: Dark blue dun or honey dun if only a few turns can be used on light wire hooks.
Setae: 4 fibres of blue dun or rusty dun spade throat hackle.
Hook: 16 medium long, Model Perfect scale.

The wings of this pattern need not be dyed since the correct color can often be obtained from the dark centers of badger or furnace hackles.

BLUE-WINGED SULPHUR DUN

Wing: Pale blue dun cut and shaped hackle point, often found in the webby centers of natural-colored hackle.
Body and Thorax: Strong sulphury yellow spun fur.
Setae: Three fibres of palest blue dun.
Hackle: Palest blue dun or cream of the hook is light and strong and can be supported with 3 or 4 turns of hackle.
Hook: 16 medium long, Model Perfect scale.

PALE SULPHUR DUN

Wings: Bluish-white cut and shaped hackle points often found in natural grey hackle.
Body and Thorax: Pale suphury yellow spun fur.

258

Setae: Three fibres of pale blue dun or palest cream.

Hackle: Pale blue dun or a few turns of cream with short fibres.

Hook: 18 medium long, Model Perfect scale.

DARK OLIVE DUN

Wings: Dark-blue dun cut and shaped hackle points.

Body and Thorax: Greenish olive spun fur, tied thinly.

Hackle: 2½ turns of honey hackle, turned in ribbing style.

Setae: Two fibres of pale blue dun.

Hook: 22 regular shank, Model Perfect scale.

LIGHT OLIVE DUN

Wings: Light blue dun cut and shaped hackle points.

Body and Thorax: Yellow olive spun fur.

Hackle: 2½ turns honey or cream hackle as for a rib.

Setae: 2 fibres of pale blue dun.

Hook: 22 regular, Model Perfect scale.

TO DRESS A QUILL-BODIED SPINNER, ANY SIZE

Prepare a porcupine quill as follows: cut off from either end, according to the color desired, enough to equal the afterbody of the natural spinner plus an extra ⅛ inch for anchorage. Clear out the soft, pithy interior; for this purpose I have found nothing better than the broken end of a very fine jeweler's-saw blade. Insert the broken end of the blade into the quill and roll the quill with thumb and forefinger. This rolling motion will cause the inner walls of the quill to bear against the fine teeth of the blade, which will scrape them clean. The blade is a better tool than a twist drill.

TYING SILK

BODY AND TAIL
ASSEMBLY

RABBIT
WHISKERS

TYING SILK

PORCUPINE
QUILL

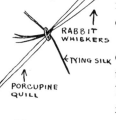

LASH ON
QUILL- TIE
IN HACKLES
AND WRAP

CLIP TOP AND
BOTTOM

Lay the open end of the prepared quill on the rear part of the shank near the bend and tie it down securely with the unwaxed tying silk which may be as fine as 00000. The quill will yield beautifully and crush tightly against the hook shank. Select 2 or 3 hackles, and tie in the butts all together at the bend of the hook, tips pointing to the rear. The number of hackles used depends on their quality and length. Wonderful results can be obtained with the clear, glassy tips of hackle which has a lot of web, if 4 or 5 of them are used. Long-fibred hackle is indicated in order to imitate the considerable spread of the wings.

Turn the hackles separately and tie off at the head to cover the thorax with densely packed turns to form a thick palmer. Now cut away all of the hackle above and below the shank, but leaving a flat wide wing of hackle fibres flaring away on either side of the thorax. The plane of hackle fibres can be left at any angle desired, either full-spent like the wings of an airplane or partly upright in a wide-angle V shape. On large patterns such as the Green Drake spinner some of the fibres can be left a little longer on the under side to provide some support and to keep the wings upright on the water, especially when the imitation is of a natural which does not habitually fall on its side.

The above suggested method of constructing spinner wings gives very good results, but there must be many people like myself who shudder at the prospect of wasting so much hackle fibre or using so many hackles for one fly if the wing is being made of genuine blue dun or bronze blue hackle. The same results can be obtained by a more economical method, requiring only one hackle and involving no cutting away of fibres, but it is a method which works well only for the full prone spinners and does not allow the operator to fix the wings at

260

different angles of elevation. The directions for this variation are as follows: before the butt of the single hackle is tied in at the rear of the thorax, fasten a short piece of spun fur or quill, or raffia, or any material as near the color of the thorax as possible, with a few turns of the silk on top of the hook at the extreme rear of the thorax; then fasten another similar piece on the bottom or ventral side. Next, the single hackle is tied in and the palmer is turned and finished at the eye. The two short pieces of material are then brought forward, splitting the fibres above and below, and the fibres are flattened to form the wings. The two short pieces of material are drawn and held taut and then fastened at the eye to finish off; this may be done more easily if the two sides are manipulated separately.

To attach tall fibres, hold the prepared quill in the thumb and forefinger of the left hand allowing the root end to protrude a little beyond the thumbnail. Take a piece of very fine unwaxed silk and hold one end firmly with the same fingers holding the quill. Select 3 fibres or only 2 for small flies and lay them alongside the quill and thread, but extending beyond the end of the quill for the proper length. Pinch all of them together firmly with the thumbnail of the left hand against the root end of the quill. Take the other end of the silk and make 5 or 6 turns around the root and the fibres. Then without slackening the pressure of the left thumbnail make several half hitches to hold the assembly, then saturate it with a drop of varnish and finally cut away all loose ends of the silk and the short butts of the tail fibres. If the brown or sharp end of the quill is being used, it is necessary to dent or score the end of the quill with the thumbnail to form ridges which will hold the turns of silk more securely. I suggest that tails be attached to the quill before the quill is bound to the hook shank. In almost

WITH THUMB
AND FOREFINGER
INDENT
QUILL AND
WHISKERS
TIE OFF WITH
HALF HITCHES

261

every case the tail fibres will bend slightly in one direction or another away from the quill; and the quill can be adjusted on the hook shank until the fibres curve above the shank, then bound in that position.

It must be noted that the dressing of the spinners requires nothing more than hackle and the porcupine quill, but the reader need not be alarmed by the simplicity of these requirements. If anything, he should rejoice and be assured that the spent patterns so tied need nothing else to complete them. Though I have mentioned spun fur, nothing extra is needed to represent the thorax, since the method of turning the hackle supplies the necessary bulk on the hook shank—the many turns of the hackle rib—creating an opaque mass that simulates this part adequately. If the reader feels along with many of the English authorities, that the eyes should be represented, he can do so, but it is doubtful that anything can be gained. Only the eyes of the male dun or spinner are very prominent, but in either case I am convinced that they are too insignificant to add or detract from any pattern. Tails are insignificant, too, but I consider them extremely important in both dun and spinner as functional necessities. They have a rudder and parachute effect for both the natural and artificial, allowing either to fall in the correct position.

The wings of spinners have always been the despair of every flytier who has attempted to imitate them. Everything imaginable has been tried in an attempt to reproduce their glossy transparent appearance. The ordinary hackle-point wing used by Halford and such as is used, for example, on the spent Adams, is utterly hopeless, because it becomes thin and narrow on the first wetting and cannot be dried and reshaped without lengthy and toilsome labor. Hackle-fibre wings tied in bunches are better but not quite good enough, since they

262

are squeezed together where they are tied to the hook shank and therefore have a tendency to cling together when wetted, though they can be dried and shaped much more quickly than the hackle-point wing. Neither of these two styles is as good as a broad, flat wing shaped from a thick palmer tied the whole length of the thorax. Above all, the use of solid materials such as the so-called fish skin, cellophane, and like substances should be carefully avoided. They have never been successful and never will be, because it it almost impossible to tie even one fly with wings of this nature without the inevitable whirling and spinning effect when it is cast. The use of such materials requires the strictest application of certain aerodynamic principles which have no place in a practical system of fly tying. In any event their history shows that flies winged in this manner are poor fish takers.

It should be noted too that in the case of spinner wings I have offered alternate colors of hackle—pale-blue dun, pale-honey dun, or bronze-blue dun. On the basis of personal experience I can say for certain that this variation in shades has no bad effect on the taking qualities of the spinner patterns. The reason for suggesting the alternate shades is also based on the theory that what is needed for the spinner wings is a hackle fibre of the glassiest transparent nature; this characteristic is usually found in neck hackles of these shades—i.e., pale-honey, pale-cream, bronze-blue, and pale-blue dun—and even some lightly marked necks of the Plymouth Rock variety, which, believe it or not, are almost as good as genuine blue dun.

As everyone knows, the bronze-blue and blue dun shades are very rare and difficult to obtain, but hackles dyed in these shades are good too, providing that the dyed hackles are very clear and glossy. White hackles are quite useless for this purpose, in fact they are probably

263

the most useless of all hackles, generally being of a chalky, dull nature and soft-textured. It is far better to dye one of the pale creams or honey necks to obtain the paramount quality of translucency in the wings of spinners.

HENDRICKSON SPINNER

Wing: Palest-blue dun or bronze blue hackle, tied palmer, then shaped half-spent or partly upright.
Body: Brown end of a porcupine quill extending about ⅜ inch beyond the bend of the hook.
Setae: 3 fibres of a pale blue dun spade or throat hackle.
Hook: 18 short shank, wide gape.

BLUE-WINGED SULPHUR SPINNER

Wing: Palest-blue dun or honey dun hackle, tied palmer, then shaped half-spent.
Body: White end of the porcupine quill, extending about ⅜ inch beyond the bend of the hook.
Setae: 3 fibres of pale blue or honey dun.
Hook: 18 short shank, wide gape.

PALE SULPHUR SPINNER

Wings: Pale-blue or honey dun hackle, tied palmer, then shaped half-spent.
Body: White end of a porcupine quill, extending about ¼ inch beyond the bend of the hook.
Setae: 3 fibres of pale-blue dun or honey dun.
Hook: 20 short shank, wide gape.

264

GREEN DRAKE SPINNER, MALE

Wings: Dark-bronze blue or bronze blue and Barred Rock mixed—tied palmer, then shaped full-spent. Width of wings about 1¼ inches from tip to tip.

Body: White end of the porcupine quill extending ½ inch beyond the bend of the hook.

Setae: 3 long rabbit whiskers or the longest chocolate hackle fibres available.

Hook: 16 wide gape, short shank.

GREEN DRAKE SPINNER, FEMALE

Wings: Pale-bronze blue hackle, tied palmer, then shaped half-spent. Width of wings about 1½ inches from tip to tip.

Body: White end of a procupine quill, the thickest of the small body quills available, extending about ½ or ⅝ inch beyond the bend of the hook.

Setae: 3 brown hackle fibres.

Hook: 14 short shank.

TO DRESS A JAPANESE BEETLE OR A FLAT-WINGED FLY OF ANY SIZE

Lay 3 short strands of black ostrich herl across the middle of the shank and tie them down with figure-eight turns to form the legs; then tie in at the bend of the hook 2 or 3 short-fibred Flemish Giant hackles which have the greenish-black stripe along the center or webby part. If these are not obtainable, any dark or black hackle will do; the Flemish Giant is suggested simply because it happens to look so much like the metallic, greenish-black underside of the beetle, though I think the color makes very little difference. Turn the hackles one at a time in

265

JASSID

TIE IN HACKLE
AND WIND FRONT

SPIRAL
HACKLE
FRONT

CLIP HACKLE
TOP AND
BOTTOM
↓

JUNGLE COCK FEATHER

TIE OFF

FOR JAP BEETLE
SAME STEPS, PLUS

BLACK
OSTRICH HERL
LEGS TIED
IN FIRST

open turns, as for a ribbing hackle, and tie off at the neck. Cut away all of the fibres above the hook shank to leave a flat table; then cut away a moderate amount underneath to form an inverted V to insure stability and a low riding appearance as with the natural (this operation can be postponed until last). Select one large or 2 medium jungle cock nails and pull off all of the loose fibres until each nail is clear; then lay it over the top of the hook concave side down and fasten the stem near the eye. Tie off and add a drop of varnish.

The procedure for tying many of the minute patterns is exactly the same as for the Japanese beetle except for the different materials which are used. This method of applying the hackle is especially good for the jassids, house flies, and smuts. It allows maximum support with the fewest possible fibres; indeed with a very light-weight, finely tempered hook, a small dry fly can be floated beautifully with only 2½ turns of hackle. No more than this should be used for these very small flies.

JAPANESE BEETLE

Wing: Largest jungle cock nail or two medium-sized nails tied flat on the back.

Hackle: Black or any dark hackle put on as for a ribbing hackle, then cut away above and below.

Hook: 16 Model Perfect, regular.

JASSID

Body: Tying silk, any color.

Wings: One medium to small jungle cock nail, any color.

Hackle: 2 or 3 very small ones turned as for a ribbing hackle, any color and as short-fibred as possible.

Hook: 20 short or 22 regular, Model Perfect.

266

The procedure in tying these important flies is even more radical than anything suggested by convention. No tying silk whatsoever is used on these flies, but if the instructions are carefully followed no trouble will be experienced.

Take a strand of horsehair or nylon monofilament of the proper color and anchor it in the middle of the shank just as if it were tying silk. Do it in this fashion: hold the short end in the left hand, the long end in the right hand. Place the horsehair or nylon squarely across the shank and wind it back on itself toward the bend. When the bend is reached, lap back toward the middle but one or two turns short of the starting point. Lap another layer toward the bend, one turn short of the second layer. Now lap another turn toward the center and bring it forward two turns on the bare shank just in front of the little ball which has been constructed in this fashion. Hang the horsehair or nylon with a heavy pliers to keep it from unwinding. Select one small hackle and clean off the fluff near the butt, leaving the little knob which is generally found there. Place the butt of the hackle underneath the bare shank and inside the last half-turn; then bring over the horsehair or nylon 3 turns to anchor the butt and push the turns together to make it more secure. Let the horsehair hang with heavy pliers. Take the tip of the hackle in the pliers and wind it criss-cross fashion over the waist of the ant, bringing it forward on the last turn, and tie in the tip with the horsehair or nylon. Continue to lap toward the eye and fill it completely; then tie off with a whip finish of the horsehair or nylon itself and the fly is complete.

A tiny drop of varnish at the heart of the hackle will make everything secure. Cut off the protruding end of horsehair or nylon at the rear of the fly which was tied down at the start of operations. This fly is really a thorax pattern tied in reverse, and I see no reason why it cannot be tied with the large ball next to the eye. How would the trout know the difference? In fact I think it would balance better since the ant has no tail support and the afterbody is rather heavy. Let the reader suit himself about this. I know of no better method of tying the ant; it will appear very clearly to the fly dresser that horsehair or nylon represent very well the hard, shiny, chitinous body of the ant.

The use of horsehair for ants may have some difficulties for many people. It is an intractable material, oftentimes slipping away from the operator and unwinding in a most exasperating manner. This can be overcome by soaking it in water for a short time. Nylon works much easier and presents no difficulties whatsoever, but it does not seem to have quite as much glint or sparkle as horsehair of good quality. The difference is not so great that one should be preferred to the other. Use the nylon if it is easier to handle. The outstanding advantage of nylon is the fact that it can be obtained in different thicknesses to suit the size of fly. It will generally be found that 5X and 6X filaments are a satisfactory range of sizes, corresponding to hook sizes 16, 18, and 20. A little experiment along this line will help the reader to determine this for his own requirements. Other dressings of the black ant have been suggested and give very good results, notably the black bear's hair and black spun fur. Both of these materials give the desired opacity for this insect, similar to the natural black horsehair.

For the red ant it is necessary to use something like nylon or the clearest washed horsehair dyed golden brown to give the desired translucency. I doubt not that there are other materials which create a similar effect. If something other than horsehair or nylon is used, tying silk must be used to anchor and tie off the material. The procedure would be exactly the same as for a dun except that the wings and tails are excluded and the hackle must be clipped above and below to fashion spent wings.

BLACK ANT

Body: In two knobs, one larger than the other, with a very thin waist between them, made of natural black horsehair, the shiny guard hairs of the black bear, or black spun fur.

Hackle: Bronze-blue tied at the waist, full-spent style.

Hook: 16, 18, 20, regular shank Model Perfect scale. (There is a great variation of size in the black ant and the larger 16's and 18's are sometimes needed.)

RED ANT

Body: In two knobs, one larger than the other, with a very thin waist between them, made of the clearest, transparent horsehair or nylon dyed golden brown.

Hackle: Bronze-blue, spent style.

Hook: 20, 22, Model Perfect, regular.

Big Springs—below McCollough's